FIELDS INSTITUTE MONOGRAPHS

THE FIELDS INSTITUTE FOR RESEARCH IN MATHEMATICAL SCIENCES

Global Dynamics, Phase Space Transport, Orbits Homoclinic to Resonances, and Applications

Stephen Wiggins

American Mathematical Society
Providence, Rhode Island

The Fields Institute
for Research in Mathematical Sciences

1000836496

The Fields Institute is named in honour of the Canadian mathematician John Charles Fields (1863–1932). Fields was a remarkable man who received many honours for his scientific work, including election to the Royal Society of Canada in 1909 and to the Royal Society of London in 1913. Among other accomplishments in the service of the international mathematics community, Fields was responsible for establishing the world's most prestigious prize for mathematics research—the Fields Medal.

The Fields Institute for Research in Mathematical Sciences is supported by grants from the Ontario Ministry of Education and Training and the Natural Sciences and Engineering Research Council of Canada. The Institute is sponsored by McMaster University, the University of Toronto, and the University of Waterloo and has affiliated universities in Ontario and across Canada.

The author received support for this volume from the National Science Foundation, Grant MSS-8958344, and from the Office of Naval Research, Grant N000148J3 023.

1991 *Mathematics Subject Classification.* Primary 58Fxx

Library of Congress Cataloging-in-Publication Data

Wiggins, Stephen.
    Global dynamics, phase space transport, orbits homoclinic to resonances, and applications/Stephen Wiggins.
        p. cm.—(Fields Institute monographs, ISSN 1069-5273)
    Includes bibliographical references and index.
    ISBN 0-8218-9202-9 (acid-free paper)
    1. Differentiable dynamical systems. 2. Global analysis (Mathematics) I. Title.
II. Series.
QA614.8.W535 1993   93-36767
514'.74—dc20   CIP

Dedication

This book is dedicated to Pat Sethna, in honor of his many contributions to applied nonlinear dynamics.

Contents

List of Figures

vii

Preface

This monograph consists of a series of weekly lectures that I gave at the Fields Institute during January through March of 1993. All of the material presented here represents joint work with colleagues Tony Leonard, Dave McLaughlin, Ed Overman II, and C. Xiong, and former Caltech students Darin Beigie, Roberto Camassa, György Haller, Tasso Kaper, Gregor Kovačič, and Vered Rom-Kedar.

I would like to thank the organizers of the Dynamical Systems and Bifurcation Theory year at the Fields Institute, John Chadam, Leon Glass, Bill Langford, Jerry Marsden, and Bill Shadwick, for making it possible for me to spend time at the Fields Institute and to give these lectures. I would like to thank Wayne Nagata and Dan Rusu for providing an early draft of the lectures. I would like to also acknowledge the wonderful help and support from the Fields Institute staff, Sheri Albers, Ron Hosler, Judy Motts, and Jaqua Taylor. Their efforts, in too many areas to mention, made my visit both very pleasant and productive. My thanks to Willem Sluis and Robin Skinner of the Fields Institute for their proofreading. I am very grateful to Brenda Law of the University of Waterloo and Liz Reidt for taking my rough LaTeX files and transforming them into a beautiful, typeset quality form. Also, my thanks to Sue Embro and Sandra Valeriote of the Fields Institute for their coordinating efforts towards the final production of this document. I would like to thank my graphic artist at Caltech, Cecelia Lin, for her work on the many figures under a very tight deadline. None of these lectures would have been given without the support of my colleague, in Applied Mechanics at Caltech, Jim Knowles. Jim took over my teaching load, on top of his own, so that I could spend three months at the Fields Institute.

The research contained in this monograph has been very generously supported by the Office of Naval Research and the National Science Foundation.

The Fields Institute for Research in Mathematical Sciences is supported by the Ontario Ministry of Education and Training and the Natural Sciences and Engineering Research Council of Canada.

Steve Wiggins

CHAPTER 1

Introduction

The goal in these lectures is to describe some aspects of the global dynamics of nonlinear dynamical systems, in particular those aspects that are relevant to applications. Ideally we would like results that hold for systems having many degrees-of-freedom. The phrase "global dynamics" carries different meanings for different groups of researchers, and for this reason we want to explain what *we* mean by this phrase and what *we* feel one should strive for in seeking results that describe the "global dynamics" of nonlinear dynamical systems. In this context, also, we want to discuss some past and recent research trends in this area and, from this viewpoint, discuss where we feel interesting future progress, and problems, can be found. In other words, we begin with a rambling discussion of ideas that will be more or less related to the rest of the lectures, but might be of some interest in their own right.

The term "dynamics" refers to the orbit structure of a dynamical system and "global" describes orbits that visit some "order one" or "non-infinitesimal" region of the phase space. Often the subject of "qualitative dynamics" or "dynamical systems" is described as the study of the asymptotic behavior of dynamical systems. While this may be the belief of mathematicians, it is certainly not the belief, or the entire need, of scientists and engineers who grapple with dynamical phenomena in their work. This will be clear in the many applications that we consider throughout these lectures.

In obtaining mathematical results describing dynamics one uses *analysis*, but in describing the relationship between a given orbit and the ambient phase space or other orbits one uses *geometry*. (Note: in this chapter the phrase "analysis" refers to the subject of mathematical analysis that a mathematician learns out of books like Walter Rudins "Mathematical Analysis" or "Real and Complex Analysis".) In many instances (and in some sense, we feel in all instances) it is the geometrical description that sets the stage where the techniques of analysis can be carried out. Thus, we feel that a central theme in the study of the global dynamics of nonlinear dynamical systems is that geometry and analysis should go hand-in-hand. Now this may seem obvious to anyone acquainted with (what they feel is) the field, however, we believe that the study of the geometry of dynamical systems and the analysis of dynamical systems has developed separately over the past 15 years and, as a result, progress in our understanding of the global dynamics of nonlinear dynamical systems with several degrees-of-freedom has not been what it could be. Let us give some examples. In particular, let us consider an area where truly remarkable progress has been made over the past 30 years – the study of the dynamics of perturbations of completely integrable Hamiltonian systems.

We first set the stage. Consider a near-integrable system with Hamiltonian

$$H(I, \theta) = H_0(I) + \epsilon H_1(I, \theta, \epsilon), \qquad (I, \theta) \in B \times T^n ,$$

where B is an open ball in $\mathbb{R}^n$. The corresponding Hamiltonian vector field is given by

$$
\begin{aligned}
\dot{I} &= -\epsilon \tfrac{\partial H_1}{\partial \theta}(I, \theta, \epsilon) , \\
\dot{\theta} &= \tfrac{\partial H_0}{\partial I}(I) + \epsilon \tfrac{\partial H_1}{\partial I}(I, \theta, \epsilon) .
\end{aligned}
\qquad (1.1)
$$

Let us also assume that the Hamiltonian is analytic on the domains of interest.

When $\epsilon = 0$, the phase space is foliated by n-tori, labeled by $I = I_0 = constant$ and with trajectories on the N-tori given by

$$
\begin{aligned}
I &= I_0 = \text{constant}, \\
\theta(t) &= \Omega(I_0)t + \theta_0, \qquad\qquad \Omega(I_0) = \tfrac{\partial H_0}{\partial I}(I_0) .
\end{aligned}
$$

First question, for $\epsilon \neq 0$ small, do any of these tori survive? This was answered by Kolmogorov [1954], Arnold [1963] and Moser [1962] in the celebrated KAM theorem. They showed that under certain nondegeneracy assumptions, the nonresonant tori, i.e., $I = I_0$ such that

$$\langle k, \Omega(I_0) \rangle = \sum_{i=1}^n k_i \Omega_i(I_0) \neq 0 \qquad \text{for all } k \in \mathbb{Z}^n - \{0\} ,$$

(note that in this case, the trajectories densely fill out the torus), survive for $\epsilon \neq 0$ if $\Omega(I_0)$ satisfies

$$|\langle k, \Omega(I_0) \rangle| \geq \frac{\delta}{|k|^n} \qquad \text{for all } k \in \mathbb{Z}^n - \{0\} .$$

What about the *resonant tori*, i.e., $I = I_0$ such that $\Omega(I_0) \neq 0$ and there is some $k^0 \neq 0$ such that

$$\langle k^0, \Omega(I_0) \rangle = 0 ?$$

It can be shown that these will also survive for $\epsilon \neq 0$ under certain nondegeneracy conditions. A key aspect of the nature of resonant tori is embodied in the idea of the *multiplicity of a resonance*. Roughly speaking, this is the number of independent integer vectors such that $\langle k, \Omega(I_0) \rangle = 0$. Multiplicity 0 corresponds to nonresonant tori, and multiplicity $n - 1$ corresponds to periodic orbits (i.e., the n-torus $I = I_0$ is foliated by periodic orbits). For multiplicities strictly between 0 and $n - 1$, there were few results, until recently. A resonance of multiplicity $m < n$ implies that the n-torus $I = I_0$ is foliated by tori of dimension $n - m$, and trajectories densely fill out these lower-dimensional tori (assuming a nondegeneracy condition $\det(\partial^2 H_0/\partial I)(I_0) \neq 0$). Recent results of Eliasson [1988], and Pöschel [1989], prove the persistence of lower-dimensional "elliptic" tori, and results of de la Llave and Wayne [1990], and Treschev [1991] prove the persistence of lower-dimensional "hyperbolic" (or "whiskered") tori. These results give us a pretty good picture of the persistence of periodic and quasiperiodic motions in perturbations of completely integrable Hamiltonian systems of the form (1.1). These tori are good examples of *global structures* in the phase space of Hamiltonian dynamical systems. What about the stable and unstable manifolds of the whiskered tori?

Let us consider another issue: As parameters are varied in numerical simulations of two-dimensional area preserving maps (such as the standard map – see Lichtenberg and Lieberman [1982], for example), some KAM tori seem to break up into "Cantori" (quasiperiodic orbits that fill out a Cantor set rather than a torus). A rigorous theory of Cantori exists only for two-dimensional area preserving maps (two-degree-of-freedom Hamiltonian systems). However there is a specific example for a four-dimensional symplectic map (coupled sawtooth maps) in a "far from integrable" setting (MacKay and Meiss [1992]). Even though this term is not often used in the this setting, we are describing a type of *global bifurcation* of KAM tori. Indeed, it is quite surprising that bifurcation theory is a very much undeveloped subject in the Hamiltonian setting. Over the past 5 years there has been progress in the study of bifurcations of equilibria and periodic orbits in Hamiltonian systems (see, e.g., Dellnitz *et al.* [1992], Golubitsky and Stewart [1987] and Meyer [1974] and references therein). However, there are virtually no results concerning the bifurcation of more complicated invariant sets (e.g., tori) in Hamiltonian systems. Why is this? One reason is that bifurcations generally occur when there is a breakdown in hyperbolicity or *normal hyperbolicity*, and there are "numbers" that measure the strength of this normal hyperbolicity–eigenvalues for equilibria, floquet multipliers for periodic orbits, and generalized Lyapunov type numbers for arbitrary normally hyperbolic invariant manifolds. Moreover, there is a theory for the persistence of general normally hyperbolic invariant manifolds (Fenichel [1971], [1974], [1977], Hirsch, Pugh, and Shub [1977], Wiggins [1993]). KAM tori are prime examples of *normally elliptic* invariant manifolds and there is no general theory for the persistence of normally elliptic invariant manifolds. Of course, the KAM theorem is a theorem that describes the persistence of certain types of normally elliptic n-tori, we will comment more on this shortly. But note that measures of normal hyperbolicity–generalized Lyapunov type numbers–provide no useful information about the persistence of KAM tori. However, for normally hyperbolic invariant manifolds they are the key quantities that allow one to construct a contraction mapping argument to prove existence and persistence under perturbation.

Let us look at one of the other great theorems of perturbations of completely integrable Hamiltonian systems– Nekhoroshev's theorem (Nekhoroshev [1971], Lochak [1992], Pöschel [1993]). This theorem refers to perturbed systems of the form of (1.1) and describes a *finite time result* for the time evolution of the actions (I variables) from their initial values:

$$|I(t) - I(0)| \le R_* \epsilon^b \quad \text{for time scales} \quad |t| \le T_* \exp(\epsilon^{-a}) \,,$$

where a, b are "stability exponents", e.g., $a = b = 1/2n$ (Lochak and Neishtadt [1992] and Pöschel [1993]).

But what about stability of specific solutions, e.g., equilibria and periodic orbits? (By "stability" we will mean an infinite time and a nonlinear result.) As a result of Liouville's theorem on the preservation of phase space volume under the Hamiltonian flow, if an eigenvalue associated with the linearization has a negative real part, there must be another eigenvalue having positive real part of the same magnitude. Hence, by the stable and unstable manifold theorem, the equilibrium is unstable. The less obvious case is the case of an elliptic equilibrium, i.e, all eigenvalues on the imaginary axis. For elliptic equilibria, with definite Hamiltonians, (the second derivative matrix of the Hamiltonian at the equilibrium is definite), locally the surfaces of constant energy look like ellipsoids shrinking down to the equilibrium which "trap" trajectories so they must remain near the equilibrium, hence we have Liapunov stability (this is made precise by the Lagrange-Dirichlet theorem). For elliptic equilibria with indefinite Hamiltonians, one can use KAM theory to obtain trapping tori, and hence Liapunov stability, in two-degree-of-freedom systems. There is also stability for non-strongly resonant equilibria $(k_1\omega_1 + k_2\omega_2 \neq 0, |k_1| + |k_2| \geq 4$, where ω_i are the imaginary parts of the eigenvalues). The stability of elliptic equilibria, with indefinite Hamiltonians in higher degree-of-freedom systems, is a largely open question. A good survey for all of these stability questions is Arnold *et al.* [1988].

For elliptic periodic and quasiperiodic orbits in systems with three or more-degrees-of-freedom, there are essentially no Liapunov or orbital stability results (except for relative equilibria of symmetric systems). However, Nekhoroshev-type estimates exist. Let us elaborate on this issue a bit more. Douady and Le Calvez [1983] have constructed an example of a four-dimensional symplectic map having an elliptic fixed point that is unstable. However, their example is a C^∞ example, and the techniques they used simply do not work in the analytic setting. (Think of this four-dimensional symplectic map as the Poincaré map arising near a periodic orbit of a three degree-of-freedom Hamiltonian system.)

What about the magical Arnold diffusion (Arnold [1964])? (It is not traditional to explain precisely what this is–but if you have time, the paper of Chierchia and Gallavotti [1992] is a rigorous study of some aspects of this phenomena.) Doesn't this immediately imply that "everything" in non-integrable Hamiltonian systems with three or more degrees-of-freedom is unstable? Who knows–for sure?

What is the reason for the lack of stability results in Hamiltonian systems? The primitive idea for proving stability is to find some "trapping" structures, such as the ellipsoids that arise in Liapunov's direct method. By "trapping structure" we mean a co-dimension, one hypersurface, that separates the phase space (energy surface) into two disjoint components such that the vector field is everywhere tangent to this hypersurface, or pointing strictly inward. For general stability questions, action-angle variables and Fourier series (on which KAM analysis relies heavily) are not really appropriate for such questions. They can be singular at interesting places. Also, as an example, consider the stability of elliptic periodic orbits in a three-degree-of-freedom system. Under suitable conditions, one may reduce the problem to the study of a four-dimensional symplectic Poincaré map. The matrix associated with the linearization about the periodic orbit has all four eigenvalues on the unit circle. If the matrix is semi-simple, the linearized map in complex coordinates takes the form

$$(z_1, z_2) \mapsto (\lambda_1 z_1, \lambda_2 z_2),$$

where $|\lambda_1| = 1 = |\lambda_2|$. The sets $|z_1|^2 + |z_2|^2 = c$ are invariant under the linearized map, and form trapping spheres for the linearized map. Would these persist for a nonlinear map? We could try a KAM-type approach, but Fourier series are appropriate for tori, not spheres. New tools are needed, which combine geometry and analysis.

Now let us go back to this question of geometry and analysis. All of the results mentioned above (other than the stability of equilibria) rely on the perturbed Hamiltonian system being expressed in the action-angle variables of the unperturbed, completely integrable system. Geometrically, this means that we are essentially considering the perturbation of tori. From the point of view of manifolds, these are particularly simple, because for practical purposes these manifolds (the tori) can be represented globally as graphs. Moreover, the tori are the ideal objects for the methods of Fourier analysis. There is another interesting characteristic about these perturbation methods for tori. As opposed to the persistence theory for normally hyperbolic invariant manifolds, the dynamics *on* the persistent tori are "qualitatively the same" (see the references to make this more precise) as that of the unperturbed tori. This is dictated by the analytical methods, which give rise to small divisor problems, which, in turn, arose due to the geometry of the tori. An important area to think about, that should lead to increased understanding in our understanding of the global *dynamics* of Hamiltonian systems is the development of a general theory of the existence and persistence of *normally elliptic* invariant manifolds. A theory where the underlying invariant manifold is more general (or at least different) than a torus.

Returning to an earlier statement, why would one claim that the study of the geometry of dynamical systems and the analysis of dynamical systems has largely proceeded separately over the past 15 years? The study of the geometry of Hamiltonian systems–symplectic structures, Lie-Poisson structures, momentum maps, the "bundleization" of mechanics, etc. – has been pursued with great vigor (Abraham and Marsden [1978], and Marsden [1992] are good sources for this point of view). However, this machinery has yielded virtually nothing in the way of results about the global dynamics of Hamiltonian systems–and certainly nothing of the magnitude of the KAM theorem, Nekhoroshev's theorem, or its descendants. (Recently, it has given rise to some results on stability and bifurcation of relative equilibria, but from the point of view of analysis, these are really local problems. (See Marsden [1992] for a review.) One might put up an argument here that much of this abstract machinery has been used in the solution of the Arnold conjecture, but in some sense, these powerful methods only tell us about periodic orbits –and nothing about stability. [1] On the other hand, many of the analytical results on the global dynamics of Hamiltonian systems has been tied to a very particular coordinate representation of the vector field–action-angle variables. One might speculate on

---

[1]The Arnold Conjecture–the number of fixed points of an exact symplectic diffeomorphism on a symplectic manifold M can be estimated below by the sum of the Betti numbers provided that the fixed points are nondegenerate. This was proved by Conley and Zehnder [1983] for $M = T^{2n}$, see Salamon [1990] for a review.

the progress that could be made if this coordinate free machinery of Hamiltonian mechanics could be joined with methods of mathematical analysis much like those developed in the context of KAM theory and Nekhoroshev's theorem.

What about applications? The results above tell us about specific geometrical objects that can occur in the phase space of a Hamiltonian dynamical system. They also give us some idea how motion occurs near these geometrical objects. Thinking in terms of analogies, one can consider these results as supplying a dynamical vocabulary. Specific applications require one to use this vocabulary to tell a story. It is important to realize that the "plot" of the story is dictated by the application–not the vocabulary. The reader shoud keep this in mind when considering the applications to come. Dynamics has always been a fertile source for the creation of new mathematics–it is important to listen to the application with good mathematical ears and hear what it has to say.

Before beginning the main material, we make a technical announcement.

Smoothness Assumptions

We will be making the blanket assumption that all vector fields and maps under consideration are sufficiently differentiable for our purposes on the regions of interest. Generally, C^r, $r \geq 2$ is sufficient. In applying KAM type results, C^r, $r \geq 2n+2$ is sufficient, where n is the number of degrees-of-freedom. In fact, all of our explicit examples will be analytic. More precise statements concerning regularity requirements can be found in the references to the specific material in each chapter.

CHAPTER 2

Homoclinic Tangles and Transport in Two-Dimensional, Time-Periodic Vector Fields

2.1.   Introduction

In this chapter we develop some aspects of the dynamics associated with homo-clinic orbits in two-dimensional time-periodic vector fields, or, equivalently, two-dimensional maps. The material from this chapter should be somewhat familiar. However, in the context of this material, we will point out the general *ideas and approaches* that will enable us to tackle the more complicated situations that arise later on.

2.2.   Melnikov's Method for Time-Periodically Forced One Degree-of-Freedom Hamiltonian Systems

Let us begin by developing a global perturbative tool, the classical Melnikov method (Melnikov [1963]), that will enable to get our "foot in the door" in re-lation to some global issues associated with homoclinic orbits. We consider the periodically forced system

$$\dot{q} = JDH(q) + \epsilon g(q, t, \epsilon), \qquad q \in \mathbb{R}^2 , \tag{2.1}$$

where

$$J = \begin{pmatrix} 0 & 1 \\ -1 & 0 \end{pmatrix}, \qquad DH(q) = \begin{pmatrix} D_1 H(q) \\ D_2 H(q) \end{pmatrix},$$

and g is periodic in t with period $T = 2\pi/\omega$. The system may contain external parameters; however, we will not display these explicitly in the notation unless they play a direct role in the immediate arguments. When $\epsilon = 0$, the unperturbed system

$$\dot{q} = JDH(q) \tag{2.2}$$

is integrable, since it is a one-degree-of-freedom Hamiltonian system.

We assume that the unperturbed system (2.2) has a hyperbolic saddle point p_0, connected to itself by a homoclinic orbit $q_h(t)$. We let $\Gamma_{p_0} = W^s(p_0) \cap W^u(p_0)$ denote the homoclinic manifold (see Figure 2.1).

We rewrite (2.1) as an autonomous system in three dimensions,

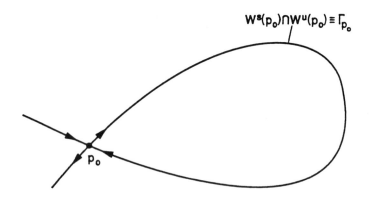

Figure 2.1. Unperturbed homoclinic manifold in the phase plane.

$$\dot{q} = JDH(q) + \epsilon g(q, \phi/\omega, \epsilon), \qquad (2.3)$$
$$\dot{\phi} = \omega, \qquad (2.4)$$

where $\phi \in S^1 = \mathbb{R}/(2\pi\mathbb{Z})$. When $\epsilon = 0$, the hyperbolic fixed point p_0 for (2.2) corresponds to the hyperbolic periodic orbit

$$\gamma(t) = (p_0, \ \phi_0 + \omega t)$$

for (2.4) in $\mathbb{R}^2 \times S^1$. The two-dimensional stable and unstable manifolds of $\gamma(t)$, $W^s(\gamma(t))$, and $W^u(\gamma(t))$ intersect along a two-dimensional homoclinic manifold $\Gamma_\gamma = \Gamma_{p_0} \times S^1$ (see Figure 2.2).

Consequences from Perturbation Theory for Normally Hyperbolic Invariant Manifolds

From the general persistence theory for normally hyperbolic invariant manifolds, along with their stable and unstable manifolds (Fenichel [1971], [1974], [1977]), when ϵ is sufficiently small, the hyperbolic periodic orbit $\gamma(t)$, along with its stable and unstable manifolds $W^s(\gamma(t))$ and $W^u(\gamma(t))$, persists as

$$\gamma_\epsilon(t) = \gamma(t) + \mathcal{O}(\epsilon) = (p_0 + \mathcal{O}(\epsilon), \ \phi_0 + \omega t),$$

along with $W^s(\gamma_\epsilon(t))$ and $W^u(\gamma_\epsilon(t))$. The terms $\mathcal{O}(\epsilon)$ in the above expression are T-periodic in t (see Figure 2.3).

The Poincaré Map

For two-dimensional time-periodic vector fields it is geometrically more convenient to consider the dynamics in the context of an associated two-dimensional Poincaré map. The construction is standard. For $\phi_0 \in S^1$, we define the global two-dimensional cross-section of the phase space $\mathbb{R} \times S^1$:

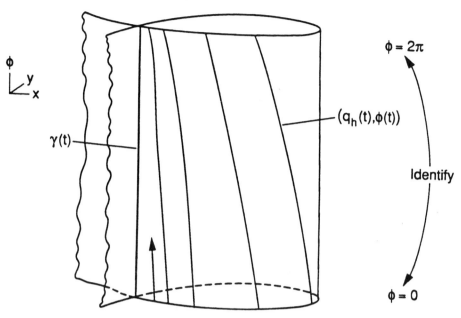

Figure 2.2. Unperturbed homoclinic manifold in the extended phase space.

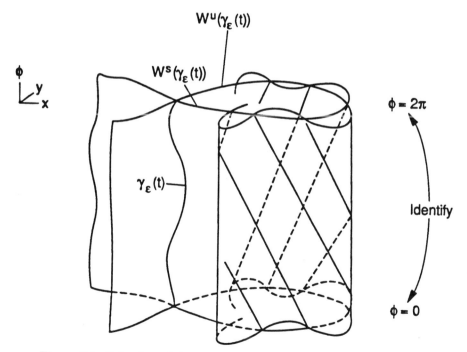

Figure 2.3. Behavior of the homoclinic manifold under perturbation.

$$\Sigma^{\phi_0} = \{(q, \phi): \ q \in \mathbb{R}^2, \ \phi = \phi_0\} \, ,$$

and construct a map of this cross-section into itself from the trajectories of the continuous time system:

$$P_\epsilon^{\phi_0} : \Sigma^{\phi_0} \mapsto \Sigma^{\phi_0} \, , \quad q_\epsilon(0) \mapsto q_\epsilon(T) \, ,$$

where $(q_\epsilon(t), \ \phi_0 + \omega t)$ is the solution of (2.3) and (2.4) with initial condition $(q_\epsilon(0), \phi_0) \in \Sigma^{\phi_0}$ at $t = 0$. For the Poincaré map, the hyperbolic periodic orbit and its stable and unstable manifolds are manifested as a hyperbolic fixed point

$$p_\epsilon(\phi_0) = \gamma_\epsilon(t) \cap \Sigma^{\phi_0}$$

with stable and unstable manifolds

$$W^s(p_\epsilon(\phi_0)) = W^s(\gamma_\epsilon(t)) \cap \Sigma^{\phi_0} \, , \quad W^u(p_\epsilon(\phi_0)) = W^u(\gamma_\epsilon(t)) \cap \Sigma^{\phi_0} \, .$$

Describing the Perturbed Phase Space Structure with Respect to the Unperturbed Phase Space Structure: Homoclinic Coordinates

On Σ^{ϕ_0} we construct *homoclinic coordinates* in a neighborhood of $\Gamma_\gamma \cap \Sigma^{\phi_0}$ by using the unperturbed homoclinic orbit. Since the unperturbed system is autonomous, if $q_h(t)$ is a trajectory then so is $q_h(t - t_0)$, for any $t_0 \in \mathbb{R}$. Hence, the map

$$t_0 \mapsto q_h(-t_0), \quad t_0 \in \mathbb{R} \, ,$$

parameterizes $\Gamma_\gamma \cap \Sigma^{\phi_0}$, and $DH(q_h(-t_0))$ is a vector normal to $\Gamma_\gamma \cap \Sigma^{\phi_0}$ at the point $q_h(-t_0)$. Moreover, $\Gamma_\gamma \cap \Sigma^{\phi_0}$ intersects $DH(q_h(-t_0))$ transversely.

By the persistence of transversal intersections under perturbation, for the perturbed system $W^s(p_\epsilon(\phi_0))$ and $W^u(p_\epsilon(\phi_0))$ transversely intersect $DH(q_h(-t_0))$ in the points q_ϵ^s and q_ϵ^u, respectively. We can, therefore, define the distance between $W^s(p_\epsilon(\phi_0))$ and $W^u(p_\epsilon(\phi_0))$, as measured along $DH(q_h(-t_0))$ in Σ^{ϕ_0}, as

$$d(t_0, \phi_0, \epsilon) = \frac{\langle q_\epsilon^u - q_\epsilon^s, \ DH(q_h(-t_0)) \rangle}{\|DH(q_h(-t_0))\|} \, ,$$

where $\langle \cdot, \cdot \rangle$ is the dot product in $\mathbb{R}^2$, $\|\cdot\| = \langle \cdot, \cdot \rangle^{1/2}$. In fact, q_ϵ^s is the unique point in $W^u(p_\epsilon(\phi_0)) \cap DH(q_h(-t_0))$ that is closest to $\gamma_\epsilon(t)$ in terms of forward time, and similarly, q_ϵ^u is the unique point in $W^u(p_\epsilon(\phi_0)) \cap DH(q_h(-t_0))$ closest to $\gamma_\epsilon(t)$ in backwards time. Figure 2.4 illustrates the geometry.

Taylor expanding the distance in ϵ gives

$$d(t_0, \phi_0, \epsilon) = \frac{\epsilon M(t_0, \phi_0)}{\|DH(q_h(-t_0))\|} + \mathcal{O}(\epsilon^2) \, , \tag{2.5}$$

where

$$M(t_0, \phi_0) = \left\langle \left.\frac{\partial q_\epsilon^u}{\partial \epsilon}\right|_{\epsilon=0} - \left.\frac{\partial q_\epsilon^s}{\partial \epsilon}\right|_{\epsilon=0}, DH(q_h(-t_0)) \right\rangle$$

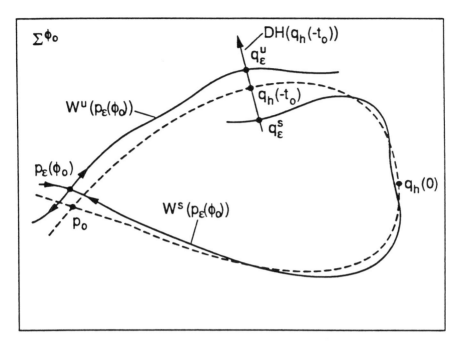

Figure 2.4. Intersection of the perturbed stable and unstable manifolds with the homoclinic coordinate system.

is defined to be the Melnikov function. It follows from (2.5) (using the implicit function theorem) that if $M(\cdot, \phi_0)$ has a simple zero at $\bar{t}_0$, then $d(\cdot, \phi_0, \epsilon)$ has a simple zero at $\bar{t}_0 + \mathcal{O}(\epsilon)$ for all sufficiently small ϵ. Hence, there exists a transverse homoclinic point to $p_\epsilon(\phi_0)$, which is $\mathcal{O}(\epsilon)$-close to $q_h(-\bar{t}_0)$.

Geometry Driven Analysis: Melnikov's Trick

We use Melnikov's trick to find a computable expression for $M(t_0, \phi_0)$. Let $q_\epsilon^s(t)$ and $q_\epsilon^u(t)$ denote the solutions of (2.1) with initial conditions q_ϵ^s and q_ϵ^u, respectively, at $t = 0$. Then put

$$M(t, t_0, \phi_0) = \left\langle \left. \frac{\partial q_\epsilon^u(t)}{\partial \epsilon} \right|_{\epsilon=0} - \left. \frac{\partial q_\epsilon^s(t)}{\partial \epsilon} \right|_{\epsilon=0} , \; DH(q_h(t - t_0)) \right\rangle ,$$

noting that $M(0, t_0, \phi_0) = M(t_0, \phi_0)$. Using the linear variational equations satisfied by

$$\left. \frac{\partial q_\epsilon^s(t)}{\partial \epsilon} \right|_{\epsilon=0} , \quad \left. \frac{\partial q_\epsilon^u(t)}{\partial \epsilon} \right|_{\epsilon=0} ,$$

Melnikov derived a first order, linear ordinary differential equation which could be simply solved to yield the Melnikov function

$$M(t_0, \phi_0) = \int_{-\infty}^{\infty} \langle\, DH(q_h(t)),\ g(q_h(t),\ (\phi_0/\omega) + t_0 + t, 0)\, \rangle\, dt\ . \qquad (2.6)$$

From (2.6), one can compute $M(t_0, \phi_0)$ without solving the perturbed system. We remark that many of the technicalities involving the definition of the distance measurement, and the subsequent derivation of the Melnikov function, are dealt with in Wiggins [1990].

Remarks

1. $M(t_0, \phi_0)$ is $(2\pi/\omega)$-periodic in t_0.

2. If $d(t_0, \phi_0, \epsilon)$ has dimensions of distance, then $M(t_0, \phi_0)$ has dimensions of flux.

3. The Melnikov function can be constructed in exactly the same way as a measure of the distance between the stable and unstable manifolds of two *different* periodic orbits, i.e., it applies to heteroclinic orbits. When this is done, one gets exactly the same result, namely, the integral in (2.6) along the unperturbed heteroclinic trajectory. The geometrical interpretation changes in the obvious ways.

Chaos

The existence of a transverse homoclinic point implies the existence of chaotic dynamics for (2.1) in the sense that, near the homoclinic point, an invariant Cantor set can be found on which the dynamics is topologically conjugate to a shift on a countable number of symbols. This follows from the Smale-Birkhoff homoclinic theorem (Wiggins [1990]).

2.3. Dynamics in Homoclinic Tangles

The existence of a transverse homoclinic point for the Poincaré map $P_\epsilon^{\phi_0}$ implies the existence of an infinite number of them, and of homoclinic *tangles*. (See Figure 2.5, and Wiggins [1990, 1992]). We examine the dynamics in tangles, and discuss the transport of phase space associated with tangles.

The Set-Up

Let $f : \mathbb{R}^2 \to \mathbb{R}^2$ be an orientation-preserving C^r diffeomorphism, $r \geq 1$, and suppose p is a hyperbolic fixed point, and q_1 is a transverse homoclinic point. Let $U(p, q_1]$ denote the segment along $W^u(p)$ with endpoints p and q_1, and similarly, let $S(p, q_1]$ denote the segment along $W^s(p)$ with endpoints p and q_1. The homoclinic point q_1 is called a *primary intersection point* or *pip* if $U(p, q_1] \cap S(p, q_1] = \{q_1\}$ (see Figure 2.5.) (Note the "open bracket" notation means that the indicated point

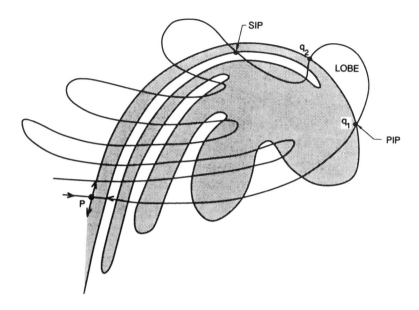

Figure 2.5. Pip's, sip's, and lobes.

adjacent to the bracket is *not* included in the segment of the manifold under consideration, the closed bracket means that it is included.) Otherwise, a homoclinic point is called a *secondary intersection point* or *sip*. (See Figure 2.5.) If q_1 and q_2 are two adjacent pip's, then the region bounded by $U[q_1, q_2] \cup S[q_1, q_2]$ is called a *lobe* (see Figure 2.5).

Below we list two important facts (Wiggins [1992]).

Facts

1. Images and pre-images of pip's are pip's.

2. Images and pre-images of lobes are lobes.

Transport and the Turnstile Mechanism

Let p be a hyperbolic fixed point, as above, and let q be a pip. Then the closed curve $U[p, q] \cup S[p, q] \equiv \mathcal{B}$ separates the phase plane $\mathbb{R}^2$ into two disjoint regions R_1 and R_2, as in Figure 2.6. We are interested in the transport of points by the map f from R_1 into R_2, and vice versa. If $S(f^{-1}(q), q) \cap U(f^{-1}(q), q)$ consists of a single pip, as in Figure 2.7, then $S[f^{-1}(q), q] \cup U[f^{-1}(q), q]$ defines exactly two lobes, denoted $L_{2,1}(1) \subset R_2$ and $L_{1,2}(1) \subset R_1$. Using the two facts given above, as well as invariance of the stable and unstable manifolds, it can be shown that $f(L_{2,1}(1)) \subset R_1$ and $f(L_{1,2}(1)) \subset R_2$ (a detailed proof can be found in Wiggins [1992]). Due to the suggestive motion of the lobes $L_{2,1}(1) \cup L_{1,2}(1)$ under iteration by f this pair of lobes is referred to as a *turnstile* (Channon and Lebowitz [1980], Bartlett [1982], MacKay, Meiss, and Percival [1984]). (Generally, the notation

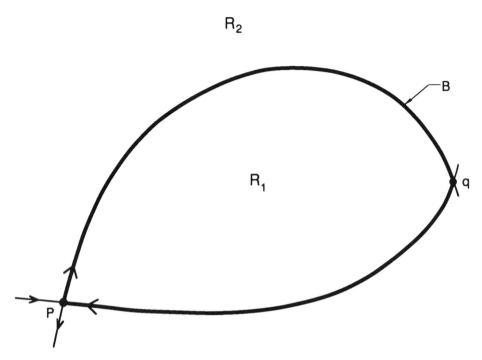

Figure 2.6. The curve $\mathcal{B}$.

$L_{i,j}(n)$ denotes a lobe consisting of points in R_i that are mapped into R_j by n iterations of f.)

In fact, one can prove the following:

Fact

The only points that move from R_1 to R_2 across the boundary $U[p,q] \cup S[p,q]$, in one iterate, are those in $L_{1,2}(1)$. Similarly, the only points that move from R_2 to R_1 across $U[p,q] \cup S[p,q]$, in one iterate, are those in $L_{2,1}(1)$.

As a consequence of this fact, all points that move from R_1 and R_2 must pass through the turnstile $L_{1,2}(1) \cup L_{2,1}(1)$. Hence, one can also show the following:

Fact

The only points that move from R_1 to R_2 across the boundary $U[p,q] \cup S[p,q]$ on the nth iterate of f are those in $L_{1,2}(1)$ on the $(n-1)^{th}$ iterate. Similarly, the only points that move from R_2 to R_1 across $U[p,q] \cup S[p,q]$ on the nth iterate of f are those in $L_{2,1}(1)$ on the $(n-1)^{th}$ iterate. (We note that multi-lobe turnstiles, self-intersecting turnstiles, and pathologies concerning lobe definition are treated in Wiggins [1992].)

Flux

From the above arguments, the area of phase space that moves from R_1 to R_2 in one iterate is

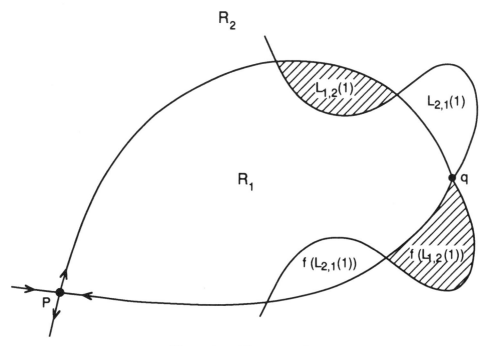

R_2

$L_{1,2}(1)$

$L_{2,1}(1)$

R_1

q

$f(L_{2,1}(1))$

$f(L_{1,2}(1))$

P

Figure 2.7. The turnstile.

$$\mu(f(L_{1,2}(1))) \, ,$$

where $\mu(A)$ is the area (two-dimensional Lebesgue measure) of A.

The Melnikov function can be used to approximate the area of a lobe/indexarea of a lobe, by the formula

$$\mu(L) \; = \; \epsilon \int_{t_{01}}^{t_{02}} |M(t_0, \phi_0)| dt_0 + \mathcal{O}(\epsilon^2) \, , \tag{2.7}$$

where L is the lobe defined by pip's at $q_h(-t_{01}) + \mathcal{O}(\epsilon)$ and $q_h(-t_{02}) + \mathcal{O}(\epsilon)$.

Formulation of a General Transport Problem for Two-Dimensional Maps

In this section we show how one can generalize these notions. Suppose $\mathbb{R}^2$ is divided up into N_R disjoint regions, denoted R_i, $i = 1, \ldots, N_R$, where the boundary of a region is made up of segments of stable and unstable manifolds of hyperbolic periodic points (or, without loss of generality, hyperbolic fixed points). Suppose that initially particles of species S_i are uniformly distributed throughout region R_i, $i = 1, \ldots, N_R$. In Figure 2.8 we illustrate the geometry.

> **Problem**: Compute the area occupied by species S_i in region R_j (for any i and j) at any specified later time.

Let us comment on this notion of the "species of a point". The species of a point indicates the region in which the point is located initially, i.e. at $t = 0$. In many transport applications it is important to know at any time $t > 0$ where a given point was located initially. An example of this would be in the study of chaotic

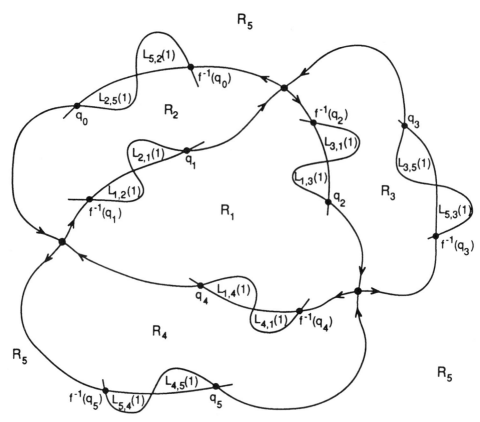

Figure 2.8. The partition of phase space into regions and the turnstiles associated with the region boundary components $q_0, \cdots, q_5$ are pip's that are the endpoints of segments of stable and unstable manifolds of hyperbolic points that are used to form segments of the boundaries of the regions. The turnstiles are then constructed as described.

advection of fluids (see Ottino [1989]) where the fluid mechanical Lagrangian point of view is essential. The general theory developed in this section incorporates this information.

We remark that in practice it is highly problem-dependent how one divides the phase space into disjoint regions separated by pieces of stable and unstable manifolds of hyperbolic periodic points. For example, in the study of transport in two-degree-of-freedom Hamiltonian systems, one might choose a finite collection of resonance bands (Mackay, Meiss, and Percival [1987]), and in the study of chaotic advection one might choose regions of the flow separated by stable and unstable manifolds of certain "stagnation points" in the flow (Rom-Kedar, Leonard, and Wiggins [1990]). In any case, it is important to realize that the only way that points can move between such regions is if they are in the lobes defined by the tangling of the stable and unstable manifolds which bound the regions. Using this observation, Rom-Kedar and Wiggins [1990] developed a theory of transport between the regions that is expressed entirely in terms of the dynamics of the turnstile lobes that control access to the different regions. Below we describe the main results, and in the next chapter we describe an application of these results.

The Main Results

We now summarize the main results of the transport theory for two-dimensional maps developed by Rom-Kedar and Wiggins [1990]. (Note: here we only give results for area preserving maps since all of our applications will be concerned with Hamiltonian systems; however the theory works just as well for dissipative systems.) First we need some more notation.

$L_{k,j}(n)$ — The lobe that leaves R_k and enters R_j on iteration n.

$L_{k,j}^i(n)$ — The portion of the lobe that leaves R_k and enters R_j on iteration n that contains species S_i, i.e. $L_{k,j}(n) \cap R_i$.

> Note that by definition $f^{n-1}(L_{k,j}(n)) = L_{k,j}(1)$ and $L_{k,j}(n) \cap L_{i,m}(n) = \emptyset$ unless $k = i$ and $j = m$ since $f^{n-1}(L_{k,j}(n)) \in R_k$, $f^{n-1}(L_{i,m}(n)) \in R_i$, $f^n(L_{k,j}(n)) \in R_j$ and $f^n(L_{i,m}(n)) \in R_m$ and the regions R_k, R_i, R_j, and R_m are, by construction, disjoint. The main quantity that we can compute by knowing the dynamics of the lobes is:

$T_{i,j}(n)$ — The area occupied by species S_i in region R_j immediately after the nth iteration.

If we know $T_{i,j}(n)$ then we also know the flux of species S_i into region R_j on the nth iterate which is given as follows.

> Flux of S_i into R_j on the nth iterate = change in the amount of species S_i in R_j on the nth iterate = $T_{i,j}(n) - T_{i,j}(n-1)$.

Before giving the formula for $T_{i,j}(n)$ derived by Rom-Kedar and Wiggins [1990] using the lobe dynamics we want to make a comment regarding the idea of flux. Recall the discussion of the motion of the lobes $L_{1,2}(1)$ and $L_{2,1}(1)$ across the boundary $\mathcal{B}$. If we are interested in the amount of R_1 that moves into R_2 on one iteration and vice versa, it should be clear that this is just the area of the lobes $L_{1,2}(1)$ and $L_{2,1}(1)$, respectively. However, the quantities $T_{i,j}(n)$ allow one to compute the long time flux of points of a specific species since at any later time a given turnstile may contain many different species. The quantity $T_{i,j}(n)$ accounts for this.

The main results found in Rom-Kedar and Wiggins [1990] are the following two formulas:

$$T_{i,j}(n) = T_{i,j}(0) + \sum_{k=1}^{N_R} \sum_{\ell=1}^{n} \left[\mu \left(L_{k,j}^i(\ell) \right) - \mu \left(L_{j,k}^i(\ell) \right) \right], \tag{2.8}$$

where $T_{i,j}(0) = 0$, $i \neq j$ and $T_{i,i}(0) = \mu(R_i)$ with the notation $\mu(A)$ denoting the area of the set A for any $A \in \mathbb{R}^2$, and

$$\mu \left(L_{k,j}^i(\ell) \right) = \sum_{s=1}^{N_R} \sum_{m=0}^{\ell-1} \mu \left(L_{k,j}(1) \cap f^m(L_{i,s}(1)) \right)$$

$$- \sum_{s=1}^{N_R} \sum_{m=1}^{\ell-1} \mu \left(L_{k,j}(1) \cap f^m(L_{s,i}(1)) \right). \tag{2.9}$$

Formulas (2.8) and (2.9) express the amount of species S_i contained in region R_j solely in terms of the dynamics of the turnstiles controlling access to the regions.

Additionally, we have the following conservation laws for the $T_{i,j}(n)$.

Conservation of Species

$$\sum_{j=1}^{N_R} (T_{i,j}(n) - T_{i,j}(n-1)) = 0 \quad , \quad i = 1, \dots, N_R \tag{2.10}$$

Conservation of Area

$$\sum_{i=1}^{N_R} (T_{i,j}(n) - T_{i,j}(n-1)) = 0 \quad , \quad j = 1, \dots, N_R. \tag{2.11}$$

CHAPTER 3

Transport in Cellular Flows

Over the past ten years much enthusiasm has arisen over the application of the methods of dynamical systems to problems concerned with mixing and transport in fluids; for a recent survey see Ottino [1989]. The general setting for these problems is as follows. Suppose one is interested in the motion of a *passive scalar* in a fluid (e.g. dye, temperature, etc.), then, *neglecting molecular diffusion*, the passive scalar follows fluid particle trajectories which are solutions of

$$\dot{x} = v(x, t; \mu), \tag{3.1}$$

where $v(x, t; \mu)$ is the velocity field of the fluid flow, $x \in \mathbb{R}^n, n = 2$ or 3, and $\mu \in \mathbb{R}^p$ represent possible parameters. When viewed as a dynamical system, the phase space of (3.1) is actually the physical space in which the fluid flow takes place. Evidently, "structures" in the phase space of (3.1) should have some influence on the transport and mixing properties of the fluid. To make this more precise, let us consider a fluid mechanically more simplified situation. Suppose that the fluid is two-dimensional, incompressible, and inviscid; then we know (Chorin and Marsden [1979]) that the velocity field can be obtained from the derivatives of a scalar valued function $\psi(x_1, x_2, t; \mu)$, known as the *stream function*, as follows:

$$
\begin{aligned}
\dot{x}_1 &= \frac{\partial \psi}{\partial x_2}(x_1, x_2, t; \mu), \\
\dot{x}_2 &= \frac{-\partial \psi}{\partial x_1}(x_1, x_2, t; \mu), & (x_1, x_2) \in \mathbb{R}^2. \quad (3.2)
\end{aligned}
$$

In the context of dynamical systems theory, (3.2) is a time-dependent Hamiltonian vector field where the stream function plays the role of the Hamiltonian. Moreover, if $\psi(x_1, x_2, t; \mu)$ depends periodically on t, then the study of (3.2) can be reduced to the study of an area preserving Poincaré map, as described in the previous chapter. In this case we would expect Smale horseshoes, resonance bands, KAM tori, and Cantori to arise in the phase space of (3.2). These structures then have a direct interpretation as actual structures in the fluid flow and they are not at all unrelated to the *coherent structures* first observed by Brown and Roshko [1974]. Furthermore, one might guess that they have an important effect on the fluid mechanics. In particular, some questions that one might ask are the following:

1. Can an understanding of the dynamics of this "structure" in the flow lead to new fluid mechanical insights?

2. Can the "structure" provide the building blocks for a simplified description of the flow?

3. Can we predict under what conditions these "structures" will be created or destroyed?

4. Can we describe the transport of fluid across such "structures" in terms of the dynamics of the "structures"?

5. Can an understanding of the dynamics of the "structures" enable us to understand the dynamics of stretching and folding of fluid line elements (i.e. interface dynamics) as a function of space and time?

In this chapter we will consider a specific example described in Camassa and Wiggins [1991] that illustrates this approach.

The Physical Setting and the Model Flow

The physical setting is as follows: We consider a (steady) convection cell whose horizontal length is much larger than its height and where the convection rolls are aligned along the y-axis (Figure 3.1). In this situation, the flow is essentially two-dimensional and, assuming stress-free boundary conditions and single-mode convection, an explicit form of the velocity field is given by (Chandrasekhar [1961])

$$\dot{x} = -\frac{A\pi}{k}\cos(\pi z)\sin(kx) = -\frac{\partial \psi}{\partial z}(x, z), \tag{3.3}$$

$$\dot{z} = A\sin(\pi z)\cos(kx) = \frac{\partial \psi}{\partial x}(x, z), \tag{3.4}$$

where $\psi(x, z) = \frac{A}{k}\sin(kx)\sin(\pi z)$, A is the maximum vertical velocity in the flow, $k = \frac{2\pi}{\lambda}$ (λ is the wavelength associated with the roll pattern), and the length measures have been non-dimensionalized so that the top is $z = 1$ and the bottom $z = 0$. This flow has a countable infinity of hyperbolic fixed points on the upper boundary at $(x, z) = (\frac{j\pi}{k}, 1)$, $j = 0, \pm 1, \pm 2, \ldots$ and a countable infinity of hyperbolic fixed points on the lower boundary at $(x, z) = (\frac{j\pi}{k}, 0)$, $j = 0, \pm 1, \pm 2, \ldots$. Fixed points with the same x coordinate are connected by a heteroclinic orbit. The result is an infinite number of cells or rolls, and many of the fluid mechanical problems of interest are concerned with the transport of a passive scalar (say, dye) from roll-to-roll. Fluid flows which are divided up into a collection of *cells* occur in a variety of applications and the methods and approaches discussed in this chapter in principle can be applied to any such *cellular flow*.

The Experiments of Solomon and Gollub

If the temperature difference between the top and bottom of the convection cell is increased, an additional time-periodic instability occurs, resulting in a time-periodic velocity field (Clever and Busse [1974], Bolton *et al.* [1986]). Solomon and Gollub [1988] experimentally studied the transport of dye in such a flow and they made the following observations:

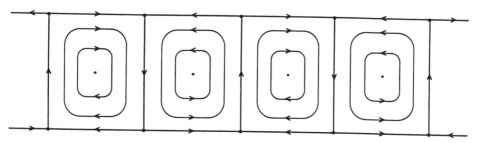

Figure 3.1. Streamlines for the steady, cellular flow.

1. There was a dramatic enhancement of the "effective diffusivity".

2. Molecular diffusion appeared to play no role in the transport.

3. The flux across the roll boundaries (heteroclinic orbits) depended linearly on the amplitude of the oscillatory instability and was independent of the wavelength λ.

Therefore, the transport problem is radically different in the unsteady case as compared to the steady case since, in the steady case, streamlines cannot cross and therefore dye crosses the roll boundaries solely as a result of molecular diffusion. In order to study these issues Solomon and Gollub [1988] introduced the following model of the *even oscillatory roll instability*:

$$\dot{x} = -\frac{A\pi}{k}\cos(\pi z)[\sin(kx) + \epsilon k f(t)\cos(kx)] = -\frac{\partial \psi}{\partial z}(x, z, t), \qquad (3.5)$$

$$\dot{z} = A\sin(\pi z)[\cos(kx) - \epsilon k f(t)\sin(kx)] = \frac{\partial \psi}{\partial x}(x, z, t), \qquad (3.6)$$

where $\psi(x, z, t) = \frac{A}{k}\sin(kx)\sin(\pi z) + \epsilon f(t)\cos(kx)\sin(\pi z)$ and we will take $f(t) = \cos(\omega t)$ and $\epsilon \sim (R - R_c)^{\frac{1}{2}}$, where R is the Rayleigh number and R_c its critical value at which the time-periodic instability occurs. The pros and cons of this model are discussed in Solomon and Gollub [1988]. This model will be the starting point of our analysis and, motivated by the experimental observations of Solomon and Gollub, we will consider the following four questions:

1. What is the mechanism for roll-to-roll transport?

2. Can we quantify the spreading of a passive scalar (dye) initially contained in one roll?

3. Can we predict the number of rolls invaded as a function of time?

4. What are the effects of the addition of a small amount of molecular diffusion?

Since the velocity field of the fluid flow is periodic in time, we will consider all of the transport issues in the context of the associated two-dimensional area-preserving Poincaré map defined on the "zero phase" cross-section. We will denote the Poincaré map generally by f.

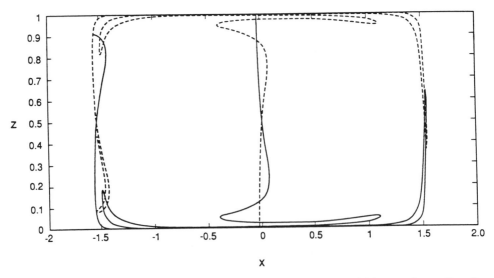

Figure 3.2. The behavior of the stable and unstable manifolds on the walls of the convection cell under the time-periodic perturbation – the heteroclinic tangle, numerically calculated for $\epsilon = 0.1$, $\omega = 0.6$, $A = 0.1$.

The Mechanism for Roll-to-Roll Transport

Under the influence of the time-periodic perturbation we expect that the heteroclinic trajectories which create the roll boundaries in the steady case to break up, giving rise to wildly oscillating lobes (Figure 3.2). This mechanism for roll-to-roll transport is completely different from that which occurs in the steady case (molecular diffusion). Melnikov's method can be used to verify this picture rigorously. In Camassa and Wiggins [1991] the Melnikov function on the zero phase cross section is calculated and found to be given by

$$M(t_0) = \omega \operatorname{sech}(\frac{\omega}{2A}) \sin(\omega t_0),$$

where $t_0 = 0$ corresponds to $z = \frac{1}{2}$. Clearly the Melnikov function has a countable infinity of simple zeros, that correspond to a countable infinity of pip's.

We label the pip's at $z = \frac{1}{2}$ by q_j and use segments of the stable and unstable manifolds of the hyperbolic fixed points on the upper and lower boundaries that end at these pip's to form roll boundaries for the steady flow as shown in Figure 3.3.

We then consider the pre-image of each pip q_j under the Poincaré map. The resulting segments of the stable and unstable manifolds between q_j and $f^{-1}(q_j)$ then form the *turnstile lobes*, which we label as described in Chapter 2. Since the Melnikov function has exactly two zeros per period it follows that each turnstile contains two lobes (Figure 3.4).

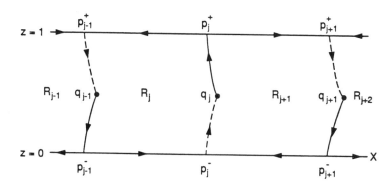

Figure 3.3. Roll boundaries for the unsteady flow (for the associated Poincaré map).

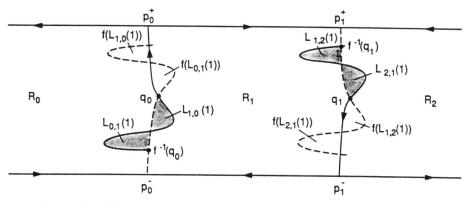

Figure 3.4. The turnstile lobes associated with roll-to-roll transport.

From the general theory described in Chapter 2, the flux of fluid between rolls is given by the area of the turnstile lobes. In this perturbative setting an approximation to the flux (lobe area) can be obtained from the Melnikov function (2.7), and is given by

$$\mu(L_{1,0}(1)) = \mu(L_{0,1}(1)) = 2\epsilon\text{sech}\frac{\omega}{2A} + \mathcal{O}(\epsilon^2).$$

We remark that because of translational symmetries, this result holds for all turnstile lobes. The symmetries of this problem are discussed in Camassa and Wiggins [1991]. This approximation to the flux shows us that the flux depends linearly on the amplitude of the oscillatory instability and is independent of the wavelength of the roll patterns λ, exactly as observed experimentally by Solomon and Gollub [1988].

The Spreading of a Passive Scalar

If we suppose the roll R_1 uniformly filled with tracer (species S_1) at $t = 0$, how much of species S_1 is contained in R_j at any $t > 0$? If we denote by $T_{1,j}(n)$ the total amount of species S_1 in R_j immediately after the n-th iterate and by $a_{1,j}(n) = T_{1,j}(n) - T_{1,j}(n-1)$ the flux of species S_1 into R_j on the n-th iterate, then it can be shown that this last formula can be reduced to an expression containing areas of intersection sets involving images of only one of the turnstile lobes $L_{1,0}(1)$ associated with the boundary of R_1. The resulting expression is given by

$$a_{1,j}(n) = T_{1,j}(n) - T_{1,j}(n-1) = (\delta_{j,2} + \delta_{j,0})\,\mu\left(L_{1,0}(1)\right)$$

$$+ \sum_{k=1}^{n-1}\left\{2\mu\left(L_{j-1,j}(1)\bigcap f^k\left(L_{1,0}(1)\right)\right) - 2\mu\left(L_{-j+2,-j}(1)\bigcap f^k\left(L_{1,0}(1)\right)\right)\right.$$

$$-2\mu\left(L_{j,j-1}(1)\bigcap f^k\left(L_{1,0}(1)\right)\right) + 2\mu\left(L_{j-1,j}(1)\bigcap f^{k-1}\left(L_{1,0}(1)\right)\right)$$

$$+\mu\left(L_{j+1,j}(1)\bigcap f^k\left(L_{1,0}(1)\right)\right) - \mu\left(L_{j,j+1}(1)\bigcap f^{k-1}\left(L_{1,0}(1)\right)\right) \qquad (3.7)$$

$$-\mu\left(L_{j,j+1}(1)\bigcap f^k\left(L_{1,0}(1)\right)\right) + \mu\left(L_{-j+1,-j}(1)\bigcap f^k\left(L_{1,0}(1)\right)\right)$$

$$+\mu\left(L_{-j+2,-j+1}(1)\bigcap f^k\left(L_{1,0}(1)\right)\right) - \mu\left(L_{j-2,j-1}(1)\bigcap f^k\left(L_{1,0}(1)\right)\right)$$

$$\left.-\mu\left(L_{j-2,j-1}(1)\bigcap f^{k-1}\left(L_{1,0}(1)\right)\right) + \mu\left(L_{j-1,j-2}(1)\bigcap f^k f\left(L_{1,0}(1)\right)\right)\right\}.$$

Consequently, a relatively straightforward numerical computation of $a_{1,j}(n)$ is possible. To compute $a_{1,j}(n)$ one merely locates the lobes indicated in (3.7), iterates $L_{1,0}(1)$ (i.e., a grid of points covering $L_{1,0}(1)$), determines the area of intersection of the appropriate iterates of $L_{1,0}(1)$ with the other lobes in (3.7), and adds up the result according to the recipe provided by (3.7). Numerically this method is extremely efficient and affords tremendous savings in CPU time over standard "brute force" methods; details are discussed in Camassa and Wiggins [1991].

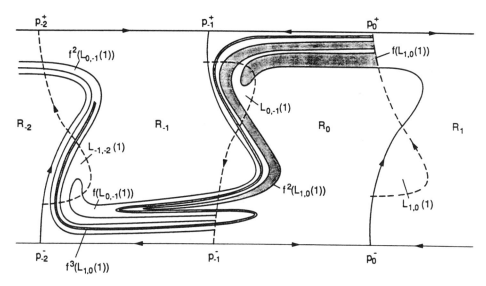

Figure 3.5. An illustration of the *signatures* of the heteroclinic tangle associated with the roll boundaries.

The Number of Rolls Invaded as a Function of Time

The answer to this question will be determined by the geometry of the lobe intersections associated with the roll boundaries. Using lobe dynamics type arguments, invariance of the manifolds as well as translation and reflection symmetries, an upper and lower bound can be placed on the time that a passive scalar initially in R_1 first invades R_j. The necessary information is contained in two integers which we refer to as the *signatures* of the heteroclinic tangle, namely:

- $\bar{m}$=the smallest integer such that $f^{\bar{m}}(L_{1,0}(1)) \cap L_{0,-1}(1) \neq \emptyset$

- $\bar{m}'$=the smallest integer such that the boundary of $f^{\bar{m}'+\bar{m}}(L_{1,0}(1))$ intersects the boundary of $L_{-1,-2}(1)$ in four distinct points.

In Figure 3.5 this is illustrated for $\bar{m} = 1$ and $\bar{m}' = 2$.

If we denote the period of the velocity field by $T = \frac{2\pi}{\omega}$ and by t_I^{-j}, the time necessary for tracer initially in R_1 to enter R_{-j}, we have the following general result obtained in Camassa and Wiggins [1991]:

$$(j\bar{m} + 1)T \leq t_I^{-j} \leq [(j-1)\bar{m}' + (\bar{m} + 1)]T \ (j \geq 2).$$

Therefore, the upper and lower bounds for the first invasion time are completely determined by the signatures $\bar{m}$ and $\bar{m}'$. This result is independent of the boundary conditions. For instance, numerical computations show that in the case $A = 0.1, \epsilon = 0.1$ and $\omega = 0.6$ we have $\bar{m} = \bar{m}' = 3$ indicating that one roll is invaded every three periods, for $A = 0.1, \epsilon = 0.1$ and $\omega = 0.24$ we have $\bar{m} = \bar{m}' = 1$ indicating that one roll is invaded every period, and for $A = 0.1, \epsilon = 0.01$ and $\omega = 0.6$ we have

$\bar{m} = \bar{m}' = 4$ indicating that one roll is invaded every four periods. We remark that the signatures are very easy to compute numerically as one needs to compute only a (usually short) finite length of the manifolds. Moreover, the necessary manifold interections are typically robust with respect to numerical errors.

Relative Time Scales of Lobe Transport Versus Transport by Molecular Diffusion

All of the transport results above neglect any possible effects of molecular diffusion. Our knowledge of the dynamics of fluid particles associated with lobes suggests the following criterion that might explain the time scale over which lobe transport dominates molecular diffusion.

The T_d ,the time scale for a tracer to diffuse across a distance of the order of a turnstile width should be long compared to the time it would take for a lobe to be mapped across the boundary of a region,i.e. T.

We have

$$T_d = \frac{(\bar{d}(\epsilon))^2}{D},$$

where $\bar{d}(\epsilon)$ is the maximum width of a turnstile lobe. Using the $\mathcal{O}(\epsilon)$ approximation to the turnstile width given by the Melnikov function, we have

$$\frac{T_d}{T} = \frac{(\epsilon \frac{\omega}{A} \mathrm{sech}(\frac{\omega}{2A}) \cosh(\frac{\pi^2 A}{2\omega})^2)}{TD}.$$

According to our criterion, lobe transport will dominate diffusive transport provided $T_d \gg T$. We check this for certain parameter values. With $D = 5.0 \times 10^{-6} \frac{cm^2}{s}$ and $A = 0.1$ we have the following three cases:

1. $\omega = 0.6, \epsilon = 0.1$ implies $T_D \simeq 200T$,

2. $\omega = 0.24, \epsilon = 0.1$ implies $T_D \simeq 300T$,

3. $\omega = 0.6, \epsilon = 0.01$ implies $T_D \simeq 2T$.

Thus, in the first two cases we would expect lobe transport to dominate molecular diffusion for about 200 and 300 periods, respectively, and in the last case only for about 2 periods. This can be checked by considering the generalized Langevin equations, which describe the behavior of molecular diffusivity:

$$\dot{x} = -\frac{\partial \psi}{\partial z} + \eta(t), \qquad (3.8)$$

$$\dot{z} = \frac{\partial \psi}{\partial x} + \zeta(t), \qquad (3.9)$$

where $\eta(t), \zeta(t)$ (diffusion terms) are random variables with a Gaussian probability distribution, such that their correlations are:

$$< \eta(t)\eta(t') > = < \zeta(t)\zeta(t') >= 2D\delta(t - t'), \qquad (3.10)$$
$$< \eta(t)\zeta(t') > = 0. \qquad (3.11)$$

In Figure 3.6 we compare the results of the lobe dynamics calculation (solid lines) and the calculations which consider molecular diffusion (dashed lines), i.e., we integrate (3.9). From this figure we can see that our criterion works reasonably well. Thus we see that geometrical and dynamical features of the lobes have led us to a physical argument that gives us a time scale on which molecular diffusion can be neglected in the process of inter-roll transport.

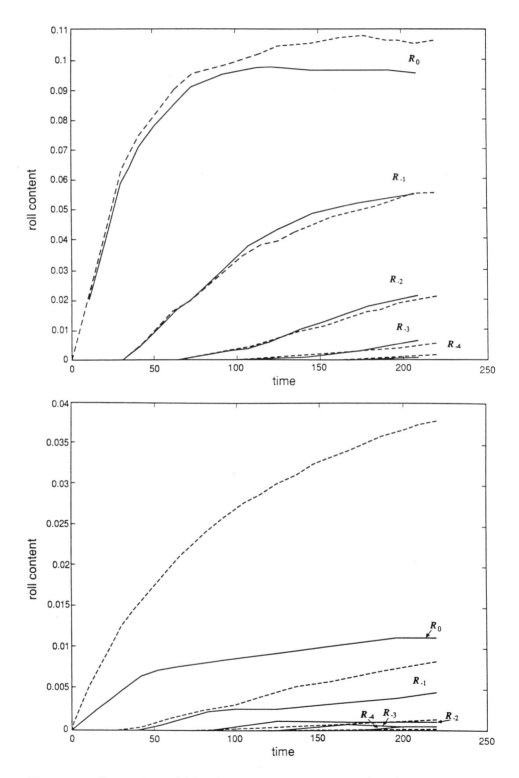

Figure 3.6. Comparison of lobe dynamic transport results (solid) with transport taking into effect molecular diffusivity (dashed) for the three cases. a) $A = 0.1$, $\omega = 0.6$, $\epsilon = 0.1$, b) $A = 0.1$, $\omega = 0.6$, $\epsilon = 0.01$.

CHAPTER 4

Homoclinic Tangles and Transport in Two-Dimensional, Time-Quasiperiodic Vector Fields

4.1. Introduction

In this chapter we show how many of the ideas from Chapter 2 can be extended to quasiperiodically time-dependent vector fields. For the sake of simplicity, we will limit our discussion to quasiperiodicity with two independent frequencies. However, the methods work equally well for an arbitrary, but finite, number of frequencies (Beigie *et al.* [1991], [1992]).

4.2. Melnikov's Method for Two-Dimensional Quasiperiodically Time-Dependent Vector Fields

We consider systems of the form

$$
\begin{aligned}
\dot{q} &= JDH(q) \ + \ \epsilon g(q, \theta_1, \theta_2, \mu, \epsilon), \\
\dot{\theta}_1 &= \omega_1, \\
\dot{\theta}_2 &= \omega_2,
\end{aligned}
\qquad (q, \theta_1, \theta_2) \in \mathbb{R}^2 \times S^1 \times S^1, \qquad (4.1)
$$

where $\mu \in \mathbb{R}^p$ represents a vector of parameters. We will see that certain aspects of the geometry of homoclinic tangles are much more sensitive to parameter variations in the quasiperiodic, as opposed to the periodic, case. A typical solution for (4.1) will be denoted by

$$
(q_\epsilon(t), \ \omega_1 t + \theta_{10}, \ \omega_2 t + \theta_{20}) \ .
$$

Assumptions on the Unperturbed Phase Space Geometry

Assume that the unperturbed system

$$
\begin{aligned}
\dot{q} &= JDH(q), \\
\dot{\theta}_1 &= \omega_1, \\
\dot{\theta}_2 &= \omega_2,
\end{aligned}
\qquad (q, \theta_1, \theta_2) \in \mathbb{R}^2 \times S^1 \times S^1, \qquad (4.2)
$$

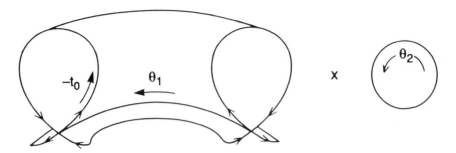

Figure 4.1. Homoclinic manifold connecting the normally hyperbolic invariant two-torus in the unperturbed problem.

has a hyperbolic fixed point p_0, connected to itself by a homoclinic orbit $q_h(t)$, i.e. $\lim_{t \to \pm\infty} q_h(t) = p_0$. Then

$$T_0 = \{(q, \theta_1, \theta_2) \mid q = p_0\}$$

is a normally hyperbolic invariant two-torus for (4.2), and

$$\Gamma_{T_0} = W^s(T_0) \cap W^u(T_0) = \{(q, \theta_1, \theta_2) \mid q = q_h(t), t \in \mathbb{R}\}$$

is a three-dimensional homoclinic manifold (Figure 4.1).

Consequences from the Perturbation Theory for Normally Hyperbolic Invariant Manifolds

From the general persistence theory for normally hyperbolic invariant manifolds, along with their stable and unstable manifolds (Fenichel [1971], [1974], [1977]), when ϵ is sufficiently small, the normally hyperbolic invariant two-torus, T_0, along with its three-dimensional stable and unstable manifolds, $W^s(T_0)$ and $W^u(T_0)$, persist, and will be denoted by T_ϵ and $W^s(T_\epsilon)$ and $W^u(T_\epsilon)$, respectively. Since the frequencies are unchanged by the perturbation, the dynamics on the torus is the same as in the unperturbed case.

The Poincaré Map

We define the three-dimensional cross-section to the vector field as follows:

$$\Sigma^{\theta_{20}} = \{(q, \theta_1, \theta_2) \mid \theta = \theta_{20}\} \ ,$$

and define the Poincaré mapping

$$P_\epsilon : \Sigma^{\theta_{20}} \to \Sigma^{\theta_{20}} \ ,$$

by

$$(q_\epsilon(0), \theta_1) \mapsto (q_\epsilon(2\pi/\omega_2), 2\pi\omega_1/\omega_2 + \theta_1) \ .$$

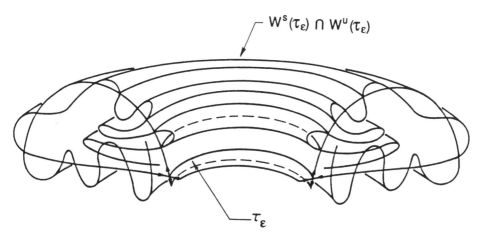

Figure 4.2. Behavior of the homoclinic manifold under perturbation (in the Poincaré section).

For the Poincaré map, $\tau_\epsilon = T_\epsilon \cap \Sigma^{\theta_{20}}$ is a normally hyperbolic invariant circle for P_ϵ, and $W^s(\tau_\epsilon)$ and $W^u(\tau_\epsilon)$ are its two-dimensional stable and unstable manifolds, respectively. However, as opposed to the unperturbed case, the two-dimensional stable and unstable manifolds of the normally hyperbolic invariant one-torus need not coincide along a branch. Rather, they may intersect transversely to form a very complex geometry structure that we illustrate in Figure 4.2.

Describing the Perturbed Phase Space Structure with Respect to the Unperturbed Phase Space Structure: Homoclinic Coordinates

On $\Sigma^{\theta_{20}}$, we parameterize $\Gamma_{\tau_0} = W^s(\tau_0) \cap W^u(\tau_0)$ by using the unperturbed trajectories as follows

$$(t_0, \theta_1) \mapsto (q_h(-t_0), \theta_1) \,,$$

and $(DH(q_h(-t_0)), 0)$ spans the one-dimensional normal space of Γ_{τ_0} at the point $(q_h(-t_0), \theta_1)$. It is easily verified that

$$W^s(\tau_0) \pitchfork (DH(q_h(-t_0)), 0), \ W^u(\tau_0) \pitchfork (DH(q_h(-t_0)), 0) \,,$$

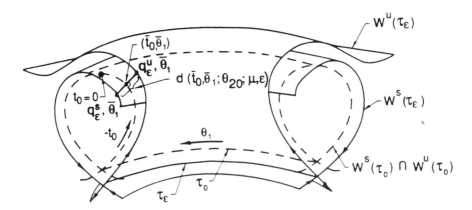

Figure 4.3. The perturbed manifolds and the homoclinic coordinates.

where the symbol $\pitchfork$ denotes transversal intersection. See Figure 4.3.

From the persistence of normally hyperbolic manifolds and of transversal intersections (restricted to compact sets), for ϵ sufficiently small, $W^s(\tau_\epsilon)$ and $W^u(\tau_\epsilon)$ intersect $(DH(q_h(-t_0)),0)$ transversally, $\mathcal{O}(\epsilon)$-close to each $(q_h(-t_0),\theta_1)$, $t_0 \in \mathbb{R}$, $\theta_1 \in S^1$, as shown in Figure 4.3.

Letting (q_ϵ^s,θ_1), (q_ϵ^u,θ_1) denote the points of transversal intersections of $W^s(\tau_\epsilon)$, $W^u(\tau_\epsilon)$ with $(DH(q_h(-t_0)),0)$, the distance between $W^s(\tau_\epsilon)$ and $W^u(\tau_\epsilon)$ at the point $(q_h(-t_0),\theta_1)$, along $(DH(q_h(-t_0)),0)$, is given by

$$d(t_0,\theta_1,\theta_{20};\mu,\epsilon) \;=\; \frac{\langle\,(DH(q_h(-t_0)),0),\ (q_\epsilon^u,\theta_1)-(q_\epsilon^s,\theta_1)\,\rangle}{\|(DH(q_h(-t_0)),0)\|}$$

$$=\; \epsilon\,\frac{M(t_0,\theta_1,\theta_{20};\mu)}{\|DH(q_h(-t_0))\|} \;+\; \mathcal{O}(\epsilon^2)\,,$$

where

$$M(t_0,\theta_1,\theta_{20};\mu) \;=\; \left\langle\, DH(q_h(-t_0)),\ \left.\frac{\partial q_\epsilon^u}{\partial\epsilon}\right|_{\epsilon=0} - \left.\frac{\partial q_\epsilon^s}{\partial\epsilon}\right|_{\epsilon=0}\,\right\rangle$$

is defined to be the *quasiperiodic Melnikov function*.

Geometry Driven Analysis: Melnikov's Trick

Using Melnikov's trick exactly as for the periodic case in Chapter 3, we obtain the formula

$$\begin{aligned}M(t_0,\theta_1,\theta_{20};\mu) \;=\;\ &\int_{-\infty}^{\infty} \langle\, xDH(q_h(t)), g(q_h(t),\ \omega_1(t+t_0)\\ &+\ \theta_1,\ \omega_2(t+t_0)+\theta_{20});\mu,0\,\rangle\ dt\,.\end{aligned}$$

Using an implicit function theorem argument, it can be shown that simple zeros of M are $\mathcal{O}(\epsilon)$-close to transversal intersections of $W^s(\tau_\epsilon)$ and $W^u(\tau_\epsilon)$.

The details can be found in Beigie *et al.* [1991] or Wiggins [1988a].

We remark that the same construction can be carried out for *heteroclinic manifolds* connecting different normally hyperbolic invariant tori. In the end, one ends up with the "same" quasiperiodic Melnikov function. The only difference is that the integrand is evaluated along the unperturbed heteroclinic orbit. The differences in geometrical interpretation are straightforward.

4.3.   The Geometry of Manifold Intersections

The geometry of manifold intersections is much richer for quasiperiodically time-dependent vector fields. We will use the quasiperiodic Melnikov function to study the geometry of manifold intersections. First we want to discuss a useful way of visualizing the manifold geometry. Imagine the following geometrical transformations:

1. Make a cut in the Poincaré section at $\theta_1 = 0$ and open up the section, identifying $\theta_1 = 0$ and $\theta_1 = 2\pi$.

2. Taking the structure obtained in the first step, cut along the normally hyperbolic one-torus and open up the resulting surface, identifying the lines where the cut was made.

3. The structure obtained from the first two steps is that of two two-dimensional surfaces "weaving in and out of each other". These are just the intersecting two-dimensional stable and unstable manifolds of the normally hyperbolic invariant one-torus. We now project this structure onto a two dimensional surface by projecting in the direction of the "thickness" of the surface.

These geometrical transformations are illustrated in Figure 4.4. As a result of this final projection, one obtains curves on a cylinder. These curves are the intersection sets of the stable and unstable manifolds of the normally hyperbolic invariant one-torus for the three-dimensional Poincaré map. This cylinder can be parametrized by $t_0 \in \mathbb{R}$ and $\theta_1 \in S^1$. Not coincidentally, these are also the independent variables of the quasiperiodic Melnikov function. In this way, the intersection sets of the manifolds can be approximated by the zero sets of the quasiperiodic Melnikov function and their relation to the geometry of the intersection of the stable and unstable manifolds can be understood through this geometrical construction.

For a general type of perturbation, the two-frequency quasiperiodic Melnikov function assumes the following functional form:

$$M(t_0, \theta_1, \theta_{20} = 0; \nu) = A_0(\mu) + A_1(\mu, \omega_1)\sin(\omega_1 t_0 + \theta_1) + A_2(\mu, \omega_2)\sin(\omega_2 t_0), \quad (4.3)$$

where $\nu \equiv (\mu, \omega_1, \omega_2)$ represent parameters. (See Beigie *et al.* [1991], [1992] for details, as well as a specific example in the next chapter.) The zero sets of this function approximate the intersection sets of the stable and unstable manifolds of

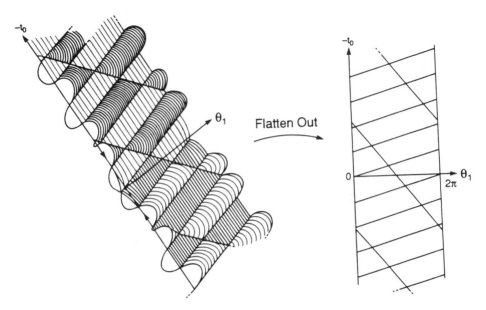

Figure 4.4. Geometrical transformations involved in visualizing manifold intersections.

the torus. Figures 4.5 (a) to (g) show the zero sets near $t_0 = 0$ for a range of A_1/A_2 values and $A_0 = 0$, $\omega_2/\omega_1 = g^{-1}$, where g is the golden mean $((\sqrt{5} - 1)/2)$. For $A_1/A_2 < 1$ the intersection manifolds are non-intersecting one-tori, for $A_1/A_2 > 1$ they are non-intersecting spirals, and for $A_1/A_2 = 1$ they are intersecting spirals (or equivalently intersecting one-tori). For other ratios of ω_2/ω_1 (both commensurate and incommensurate), the geometry of the intersection manifolds as A_1/A_2 is varied (keeping $A_0 = 0$) is qualitatively similar to the situations shown here. There is a technical point concerning the non-generic case of $A_1/A_2 = 1$: the fact that the zero sets cross at this parameter value does not necessarily imply that the intersection manifolds cross at the same parameter values; however, in Beigie et al. [1992] it is shown that for each crossing of the $M = 0$ sets, the intersection manifolds cross within $\mathcal{O}(\epsilon)$ for nearby parameter values. Figure 4.5(h) shows a case with $A_0 = \pm 1.5$, $A_1 = A_2 = \pm 1$, and $\omega_2/\omega_1 = 2$: the intersection manifolds exist for a *subset* of $\theta_1 \in [0, 2\pi)$ values. The geometry of the intersection manifolds for $A_0 \neq 0$ may be qualitatively different for commensurate and incommensurate frequencies. It is shown in Beigie et al. [1992] that in the commensurate frequency case the intersection manifolds may exist only on subsets of $\theta_1 \in [0, 2\pi)$ for particular parameter values.

4.4.   Lobes and Flux

Using the above discussion, as well as Figure 4.5, as motivation, we now discuss partitioning of the Poincaré section by segments of stable and unstable manifolds of normally hyperbolic invariant tori, as well as transport across this "dividing surface", by means of the turnstile mechanism.

Our discussion will be in a perturbative setting so that we can use the quasiperiodic Melnikov function as a tool for understanding manifold geometry. Let τ_ϵ^a and τ_ϵ^b denote two normally hyperbolic one-tori with $W^s(\tau_\epsilon^a)$, $W^u(\tau_\epsilon^a)$, $W^s(\tau_\epsilon^b)$, $W^u(\tau_\epsilon^b)$ denoting the respective two-dimensional stable and unstable manifolds. We want to consider the geometry of the intersection of $W^u(\tau_\epsilon^a)$ and $W^s(\tau_\epsilon^b)$. We first want to define the notion of a *primary intersection manifold*.

Primary Intersection Manifolds and Primary Intersection Points

Suppose that the point $(\tilde{t}_0, \tilde{\theta}_1)$ satisfies $M(\tilde{t}_0, \tilde{\theta}_1, \theta_{20}; \tilde{\nu}) = 0$ and $\frac{\partial M}{\partial t_0}(\tilde{t}_0, \tilde{\theta}_1, \theta_{20}; \tilde{\nu}) \neq 0$. Then, by the implicit function theorem, there exists a C^r function (provided the vector field is C^r) $t_0(\theta_1; \theta_{20})$ on some domain

$$\mathcal{Z} \equiv (\alpha_1, \beta_1) \in T^1,$$

where $\tilde{\theta}_1 \in (\alpha_1, \beta_1) \subset [0, 2\pi)$, such that

$$M(t_0(\theta_1; \theta_{20}), \theta_1, \theta_{20}; \tilde{\nu}) = 0$$

(for simplicity of notation the dependence of the above function t_0 on the parameters $\tilde{\nu}$ will not be explicitly included in the notation). We denote the closure of $\mathcal{Z}$ by

$$\bar{\mathcal{Z}} \equiv [\alpha_1, \beta_1] \in T^1,$$

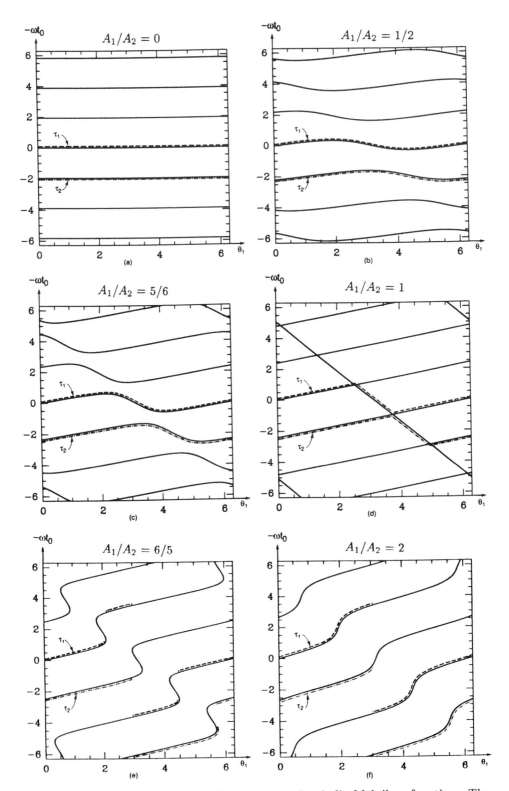

Figure 4.5. Zero sets of the two-frequency quasiperiodic Melnikov function. The dashed lines represent possible choices for primary intersection manifolds (pim's).

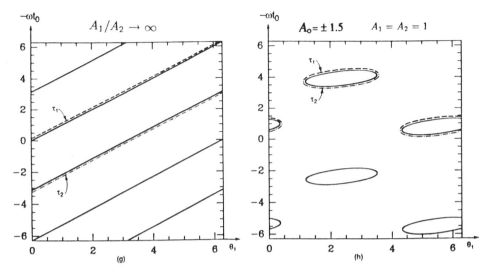

(g)   (h)

and if $\beta_1 = 2\pi$ then we define $[\alpha_1, \beta_1] = [\alpha_1, 2\pi)$.

Definition 4.4.1 (Primary Intersection Manifold). *Let* $t_0^i(\theta_1; \theta_{20})$ *with domain* $\mathcal{Z}^i$, $i = 1, \ldots, p$, *be functions as defined above with* $\mathcal{Z}^i \cap \mathcal{Z}^j = \emptyset$ *and* $\bar{\mathcal{Z}}^i \cap \bar{\mathcal{Z}}^j \equiv \mathcal{Z}^{ij}$ $\forall i, j$ $i \neq j$. *Then the set*

$$\{(t_0, \theta_1) \mid t_0 = t_0^i(\theta_1; \theta_{20}), \theta_1 \in \bar{\mathcal{Z}}^i, i = 1, \ldots, p\}$$

parametrizes a one-dimensional surface contained in $W^s(\tau_\epsilon^a) \cap W^u(\tau_\epsilon^b)$. *In order for this surface to be a single-valued graph over* $\bar{\mathcal{Z}}^1 \times \cdots \times \bar{\mathcal{Z}}^p$, *we further specify*

$$t_0 = t_0^i(\theta_1; \theta_{20}) \text{ on } \mathcal{Z}^{ij} \qquad i \neq j$$

(provided $\mathcal{Z}^{ij} \neq \emptyset$). *We refer to this surface as a primary intersection, manifold (pim), which we denote by* τ. *Points in the intersection of* $W^u(\tau_\epsilon^a)$ *and* $W^s(\tau_\epsilon^b)$ *not in primary intersection manifolds are said to be in secondary intersection manifolds (sim).*

This definition allows for several geometrical possibilities. We list some fundamental ones below, and illustrate them in Figure 4.5 where the dashed lines (denoted τ_1 and τ_2) are some possible choices for pim's.

$\boxed{p = 1 \textbf{ and } \bar{\mathcal{Z}}^1 = T^1.}$ In this case τ is either a one-torus or a one-dimensional segment of a spiral manifold. See Figures 4.5(a),(b),(c) for tori, and Figure 4.5(g) for spirals. Note that one could define spiral pim's in Figure 4.5(f) as well, but we will use this figure to illustrate another type of definition of pim's.

$\boxed{p > 1, \, \bar{\mathcal{Z}}^1 \times \cdots \times \bar{\mathcal{Z}}^p = T^1, \textbf{ and } t_0^i(\theta_1; \theta_{20}) = t_0^j(\theta_1; \theta_{20}) \textbf{ on } \mathcal{Z}^{ij}, \, i \neq j.}$ In this case τ is again either a one-torus or a one-dimensional segment of a spiral manifold. What typically happens in this situation is that $t_0^i(\theta_1; \theta_{20})$

and $t_0^j(\theta_1; \theta_{20})$ undergo a bifurcation on $\mathcal{Z}^{ij}$ and τ is formed by piecing together the functions $t_0^i(\theta_1; \theta_{20})$, $i = 1, \ldots, p$, at the surfaces where they bifurcate. (Note that the dimension of $\mathcal{Z}^{ij}$ is generically 0, and it is necessary for $D_{(t_0, \theta_1)} M(t_0, \theta_1, \theta_{20}; \nu)$ to have rank zero on these surfaces.) See Figure 4.5(d) for an example.

$\boxed{p > 1,\ \bar{\mathcal{Z}}^1 \times \cdots \times \bar{\mathcal{Z}}^p = T^1,\ \text{and } t_0^i(\theta_1; \theta_{20}) \neq t_0^j(\theta_1; \theta_{20}) \text{ on } \mathcal{Z}^{ij}, i \neq j.}$ In   this
case τ is discontinuous on the $\mathcal{Z}^{ij}$; however, τ is a single-valued graph over T^1 from our previous condition on the definition of pim's. See Figures 4.5(e) and (f) for examples.

$\boxed{p > 1 \text{ and } \mathcal{Z}^{ij} = \emptyset.}$ Unlike the previous cases, $\bar{\mathcal{Z}}^1 \times \cdots \times \bar{\mathcal{Z}}^p \subset T^{\ell-1}$ and τ contains gaps. We define a pim on a *subset* of T^1 when intersection manifolds do not cover T^1, as in Figure 4.5(h).

These four cases are not exhaustive; for example, one may have combinations of the second, third and fourth cases. Namely, τ may be nondifferentiable and/or discontinuous and/or possess gaps on $\mathcal{Z}^{ij}$.

Before using pim's to construct a "dividing surface" in the Poincaré section we need the following three concepts:

Definition 4.4.2 (Primary Intersection Points). *For $\bar{\theta}_1 \in \bar{\mathcal{Z}}^1 \times \cdots \times \bar{\mathcal{Z}}^p$, the intersection of the pim τ with the phase slice $\chi(\bar{\theta}_1) \equiv \{(x, \theta_1) | \theta_1 = \bar{\theta}_1\}$ defines a unique point. We refer to this point $p(\bar{\theta}_1) = \tau \cap \chi(\bar{\theta}_1)$ as a primary intersection point (pip). Intersection points in the phase slice that are not primary intersection points are said to be secondary intersection points (sip).*

Definition 4.4.3. *Let $p_1(\bar{\theta}_1)$ and $p_2(\bar{\theta}_1)$ denote two pip's in the phase slice $\chi(\bar{\theta}_1)$. Then let $S[p_1(\bar{\theta}_1), p_2(\bar{\theta}_1)]$ and $U[p_1(\bar{\theta}_1), p_2(\bar{\theta}_1)]$ denote the segments of $W^s(\tau_\varepsilon^a) \cap \chi(\bar{\theta}_1)$ and $W^u(\tau_\varepsilon^b) \cap \chi(\bar{\theta}_1)$, respectively, from $p_1(\bar{\theta}_1)$ to $p_2(\bar{\theta}_1)$. We say the two pip's are adjacent if $S[p_1(\bar{\theta}_1), p_2(\bar{\theta}_1)]$ and $U[p_1(\bar{\theta}_1), p_2(\bar{\theta}_1)]$ contain no other pip's besides $p_1(\bar{\theta}_1)$ and $p_2(\bar{\theta}_1)$.*

Definition 4.4.4 (Lobes). *For all $\bar{\theta}_1 \in \bar{\mathcal{Z}}^1 \times \cdots \times \bar{\mathcal{Z}}^p$, let $p_1(\bar{\theta}_1)$, $p_2(\bar{\theta}_1)$ denote adjacent pip's in the phase slice $\chi(\bar{\theta}_1)$. A lobe $\mathcal{L}$ is a 3-dimensional region in $\Sigma^{\theta_{20}}$ such that:*

(a) *for all $\bar{\theta}_1 \in \bar{\mathcal{Z}}^1 \times \cdots \times \bar{\mathcal{Z}}^p$, $\mathcal{L} \cap \chi(\bar{\theta}_1)$ is the region in $\chi(\bar{\theta}_1)$ bounded by $S[p_1(\bar{\theta}_1), p_2(\bar{\theta}_1)] \cup U[p_1(\bar{\theta}_1), p_2(\bar{\theta}_1)]$, and*

(b) *the sign of $M(t_0, \bar{\theta}_1, \theta_{20}; \nu)$ in the interval $t_0(\bar{\theta}_1) \in [t_0^1(\bar{\theta}_1), t_0^2(\bar{\theta}_1)]$, where $t_0^i(\bar{\theta}_1)$ is the t_0 value associated with $p_i(\bar{\theta}_1)$, for $i = 1, 2$, is independent of $\bar{\theta}_1 \in \bar{\mathcal{Z}}^1 \times \cdots \times \bar{\mathcal{Z}}^p$.*

Transport and Turnstiles

Let τ_c be a pim with $p_c(\theta_1) \equiv \tau_c \cap \chi(\theta_1)$ and denote $p_\epsilon^a(\theta_1) \equiv \tau_\epsilon^a \cap \chi(\theta_1)$, $p_\epsilon^b(\theta_1) \equiv \tau_\epsilon^b \cap \chi(\theta_1)$. Then

$$\mathcal{S} = \left\{ (x,y,\theta_1) \,|\, (x,y) = U[p_\epsilon^a(\theta_1), p_c(\theta_1)] \cup S[p_\epsilon^b(\theta_1), p_c(\theta_1)],\ \theta_1 \in \bar{\mathcal{Z}}^1 \times \cdots \times \bar{\mathcal{Z}}^p \right\} \tag{4.4}$$

is a 2-dimensional surface in $\Sigma^{\theta_{20}}$ which intersects each two-dimensional phase slice, $\chi(\theta_1)$ in a curve. We can view this curve as separating the phase slice into two *regions*, $R_1(\theta_1)$ and $R_2(\theta_1)$, and we want to consider the motion of points across $\mathcal{S} \cap \chi(\theta_1)$, for $\theta_1 \in \bar{\mathcal{Z}}^1 \times \cdots \times \bar{\mathcal{Z}}^p$. First, it will be useful to recall that the Poincaré map acts as follows:

$$P_\epsilon(\chi(\theta_1)) = \chi\left(\theta_1 + 2\pi \frac{\omega_1}{\omega_2}\right),$$

with

$$R_1(\theta_1) \to R_1\left(\theta_1 + 2\pi \frac{\omega_1}{\omega_2}\right),$$
$$R_2(\theta_1) \to R_2\left(\theta_1 + 2\pi \frac{\omega_1}{\omega_2}\right). \tag{4.5}$$

Now we will construct certain special lobes, called *turnstile lobes*, that mediate the transport across $\mathcal{S}$. Our construction will heavily use the quasiperiodic Melnikov function.

Choose $\theta_1 \in \bar{\mathcal{Z}}^1 \times \cdots \times \bar{\mathcal{Z}}^p$ and let $t_0^c(\theta_1)$ denote the zero of $M(t_0, \theta_1; \theta_{20})$, *restricted to* $\chi(\theta_1)$, corresponding to $\tau_c \cap \chi(\theta_1) \equiv p_c(\theta_1)$ and $t_0^{-c}(\theta_1)$ denote the zero of $M(t_0, \theta_1; \theta_{20})$, *restricted to* $\chi(\theta_1)$, corresponding to $P_\epsilon^{-1}(\tau_c) \cap \chi(\theta_1) \equiv p_c^{-1}(\theta_1)$. Next let $\mathcal{N}(\theta_1)$ be the number of zeros of $M(t_0, \theta_1; \theta_{20})$, *restricted to* $\chi(\theta_1)$, between (but not including) $t_0^c(\theta_1)$ and $t_0^{-c}(\theta_1)$. We denote these zeros by

$$t_0^{-c}(\theta_1) \equiv t_0^0(\theta_1) < t_0^1(\theta_1) < \cdots < t_0^{\mathcal{N}(\theta_1)}(\theta_1) < t_0^c(\theta_1) \equiv t_0^{\mathcal{N}(\theta_1)+1}(\theta_1).$$

Clearly, if we choose any open interval $\left(t_0^i(\theta_1), t_0^{i+1}(\theta_1)\right)$, $i = 1, \ldots, \mathcal{N}(\theta_1)$, then $M(t_0, \theta_1; \theta_{20})$ is of one sign on this interval.

Let $U[p_c^{-1}(\theta_1), p_c(\theta_1)]$ denote the segment of $W^u(\tau_\epsilon) \cap \chi(\theta_1)$ with endpoints $p_c^{-1}(\theta_1)$ and $p_c(\theta_1)$ and let $S[p_c^{-1}(\theta_1), p_c(\theta_1)]$ denote the segment of $W^s(\tau_\epsilon) \cap \chi(\theta_1)$ with endpoints $p_c^{-1}(\theta_1)$ and $p_c(\theta_1)$. Then on $\chi(\theta_1)$, $U[p_c^{-1}(\theta_1), p_c(\theta_1)]$ and $S[p_c^{-1}(\theta_1), p_c(\theta_1)]$ intersect to form two *sets* of two-dimensional lobes in the phase slice $\chi(\theta_1)$ which we denote as $L_{1,2}(1, \theta_1)$ and $L_{2,1}(1, \theta_1)$, respectively. These sets of lobes are characterized as follows:

$L_{1,2}(1, \theta_1)$ (respectively $L_{2,1}(1, \theta_1)$) is the set of lobes such that

1. $L_{1,2}(1, \theta_1) \subset R_1(\theta_1)$ (respectively $L_{2,1}(1, \theta_1) \subset R_2(\theta_1)$),

2. $M(t_0, \theta_1; \theta_{20})$, *restricted to* $\chi(\theta_1)$, is the same sign on the interval $\left(t_0^i(\theta_1), t_0^{i+1}(\theta_1)\right)$, for some $i \in \{1, \ldots, \mathcal{N}(\theta_1)\}$, where $t_0^i(\theta_1)$ and $t_0^{i+1}(\theta_1)$ correspond to the pip's defining a lobe in $L_{1,2}(1, \theta_1)$ (respectively $L_{2,1}(1, \theta_1)$).

Let $\mathcal{N}_{1,2}(1,\theta_1)$ denote the number of lobes in the set $L_{1,2}(1,\theta_1)$ and let $\mathcal{N}_{2,1}$ $(1,\theta_1)$ denote the number of lobes in the set $L_{2,1}(1,\theta_1)$. Then we have

$$L_{1,2}(1,\theta_1) \equiv L_{1,2}(1,\theta_1;1) \cup \cdots \cup L_{1,2}(1,\theta_1;\mathcal{N}_{1,2}(1,\theta_1)),$$

$$L_{2,1}(1,\theta_1) \equiv L_{2,1}(1,\theta_1;1) \cup \cdots \cup L_{2,1}(1,\theta_1;\mathcal{N}_{2,1}(1,\theta_1)),$$

$$(4.6)$$

and, clearly,

$$\mathcal{N}_{1,2}(1,\theta_1) + \mathcal{N}_{2,1}(1,\theta_1) = \mathcal{N}(\theta_1) + 1.$$

The lobes $L_{1,2}(1,\theta_1) \cup L_{2,1}(1,\theta_1)$ are the generalization to the case of quasiperiodic velocity fields of the notion of a *turnstile* that we introduced in chapter 2.

The significance of the turnstile lobes lies in the following facts proved in Beigie *et al.* [1991].

Fact 1:

$$P_\epsilon\left(L_{1,2}(1,\theta_1)\right) \subset R_2\left(\theta_1 + 2\pi\frac{\omega_1}{\omega_2}\right),$$

$$P_\epsilon\left(L_{2,1}(1,\theta_1)\right) \subset R_1\left(\theta_1 + 2\pi\frac{\omega_1}{\omega_2}\right).$$

Fact 2:

The only points that enter $R_2\left(\theta_1 + 2\pi n\frac{\omega_1}{\omega_2}\right)$ on the nth iterate of P_ϵ are those that are in $L_{1,2}(1,\theta_1 + 2\pi(n-1)\frac{\omega_1}{\omega_2})$ on the $(n-1)$-iterate of P_ϵ. Similarly, the only points that enter $R_1\left(\theta_1 + 2\pi n\frac{\omega_1}{\omega_2}\right)$ on the nth iterate of P_ϵ are those that are in $L_{2,1}(1,\theta_1 + 2\pi(n-1)\frac{\omega_1}{\omega_2})$ on the $(n-1)$-iterate of P_ϵ.

In Figure 4.6 we illustrate the turnstile construction.

From these two results we conclude that the turnstile lobes are the mediators of transport between the regions $R_1(\theta_1)$ and $R_2(\theta_1)$. Preimages of the turnstile lobes are formed in the usual way. On the phase slice $\chi(\theta_1)$ we consider $P_\epsilon^{-n}(\tau_c) \cap \chi(\theta_1) \equiv p_c^{-n}(\theta_1)$ and $P_\epsilon^{-n+1}(\tau_c) \cap \chi(\theta_1) \equiv p_c^{-(n+1)}(\theta_1)$, $(n > 1)$. Then $U[p_c^{-(n+1)}(\theta_1), p_c^{-n}(\theta_1)]$ and $S[p_c^{-(n+1)}(\theta_1), p_c^{-n}(\theta_1)]$ intersect to form two families of lobes:

$$L_{1,2}(n,\theta_1) \equiv L_{1,2}(n,\theta_1;1) \cup \cdots \cup L_{1,2}(n,\theta_1;\mathcal{N}_{1,2}(n,\theta_1)),$$

$$L_{2,1}(n,\theta_1) \equiv L_{2,1}(n,\theta_1;1) \cup \cdots \cup L_{2,1}(n,\theta_1;\mathcal{N}_{2,1}(n,\theta_1)),$$

$$(4.7)$$

and the action of the Poincaré map on these lobes is as follows:

$$P_\epsilon\left(L_{1,2}(n,\theta_1)\right) \equiv L_{1,2}(n-1,\theta_1 + 2\pi\frac{\omega_1}{\omega_2})$$

$$= L_{1,2}\left(n-1,\theta_1 + 2\pi\frac{\omega_1}{\omega_2};1\right) \cup \cdots \cup L_{1,2} \qquad (4.8)$$

$$\left(n-1,\theta_1 + 2\pi\frac{\omega_1}{\omega_2};\mathcal{N}_{1,2}(n-1,\theta_1 + 2\pi\frac{\omega_1}{\omega_2})\right),$$

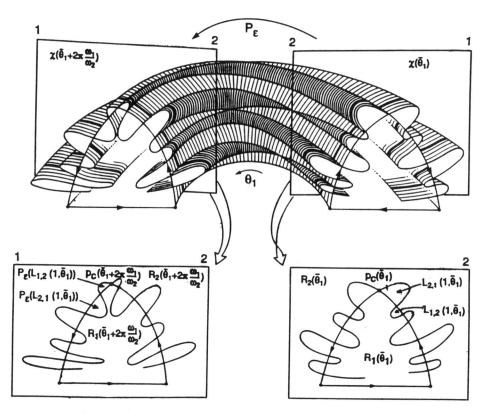

Figure 4.6. Geometry behind the turnstile construction.

$$P_\epsilon\left(L_{2,1}(n,\theta_1)\right) \equiv L_{2,1}(n-1,\theta_1+2\pi\tfrac{\omega_1}{\omega_2})$$

$$= L_{2,1}\left(n-1,\theta_1+2\pi\tfrac{\omega_1}{\omega_2};1\right) \cup \cdots \cup L_{2,1} \tag{4.9}$$

$$\left(n-1,\theta_1+2\pi\tfrac{\omega_1}{\omega_2};\mathcal{N}_{2,1}(n-1,\theta_1+2\pi\tfrac{\omega_1}{\omega_2})\right).$$

Moreover, the following relations hold:

$$\mathcal{N}_{1,2}(n,\theta_1) = \mathcal{N}_{1,2}(1,\theta_1+2\pi\tfrac{\omega_1}{\omega_2}(n-1)),$$
$$\mathcal{N}_{2,1}(n,\theta_1) = \mathcal{N}_{2,1}(1,\theta_1+2\pi\tfrac{\omega_1}{\omega_2}(n-1)), \tag{4.10}$$

which implies that the numbers $\mathcal{N}_{1,2}(n,\theta_1)$ and $\mathcal{N}_{2,1}(n,\theta_1)$ can be calculated from the properties of the zeros of the quasiperiodic Melnikov function. In Figure 4.7 we illustrate the construction of the turnstile lobes and their behavior under iteration by the Poincaré map for the general quasiperiodic Melnikov function given in (4.3). (See also Figure 4.5.)

Lobe Area and Flux

Using the turnstile construction given in the previous section, we can begin to quantify the notion of transport between $R_1(\theta_1)$ and $R_2(\theta_1)$. We first describe some notions of *flux*. The symbol $\mu(A)$ will denote the area or volume of the set A, which may either be restricted to a phase slice, or a set contained in the Poincaré section.

Instantaneous Flux

The instantaneous flux from $R_1\left(\theta_1+2\pi n\tfrac{\omega_1}{\omega_2}\right)$ into $R_2\left(\theta_1+2\pi(n+1)\tfrac{\omega_1}{\omega_2}\right)$ under iteration by P_ϵ, $\phi_{1,2}\left(\theta_1+2\pi n\tfrac{\omega_1}{\omega_2}\right)$, is given by

$$\phi_{1,2}\left(\theta_1+2\pi n\frac{\omega_1}{\omega_2}\right) = \frac{\omega_2}{2\pi}\mu\left(L_{1,2}\left(1,\theta_1+2\pi n\frac{\omega_1}{\omega_2}\right)\right).$$

Similarly, the instantaneous flux from $R_2\left(\theta_1+2\pi n\tfrac{\omega}{\omega_2}\right)$ into $R_1\left(\theta_1+2\pi\,(n+1)\tfrac{\omega}{\omega_2}\right)$ under iteration by P_ϵ, $\phi_{2,1}\left(\theta_1+2\pi n\tfrac{\omega_1}{\omega_2}\right)$, is given by

$$\phi_{2,1}\left(\theta_1+2\pi n\frac{\omega}{\omega_2}\right) = \frac{\omega_2}{2\pi}\mu\left(L_{2,1}\left(1,\theta_1+2\pi n\frac{\omega_1}{\omega_2}\right)\right).$$

Average Flux

The average flux across the surface $\mathcal{S}$, from $R_1 \equiv \cup_{\theta_1 \in \bar{z}^1 \times \cdots \times \bar{z}^p} R_1(\theta_1)$ into $R_2 \equiv \cup_{\theta_1 \in \bar{z}^1 \times \cdots \times \bar{z}^p} R_2(\theta_1)$, $\phi_{1,2}(\theta_1)$, is given by

$$\phi_{1,2}(\theta_1) = \langle \phi_{1,2}\left(\theta_1+2\pi n\frac{\omega_1}{\omega_2}\right)\rangle_n.$$

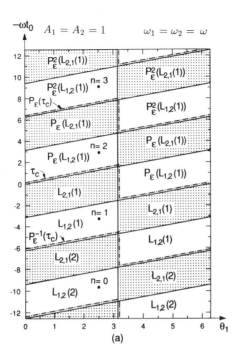

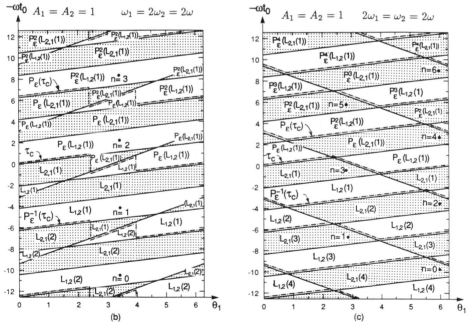

Figure 4.7. Turnstile dynamics for the various parameter values of the general quasiperiodic Melnikov function given earlier. The points labeled $n = 0, 1, 2$ are some points of the forward orbit (starting at $n = 0$) of a representative point. (*Continued on next page.*)

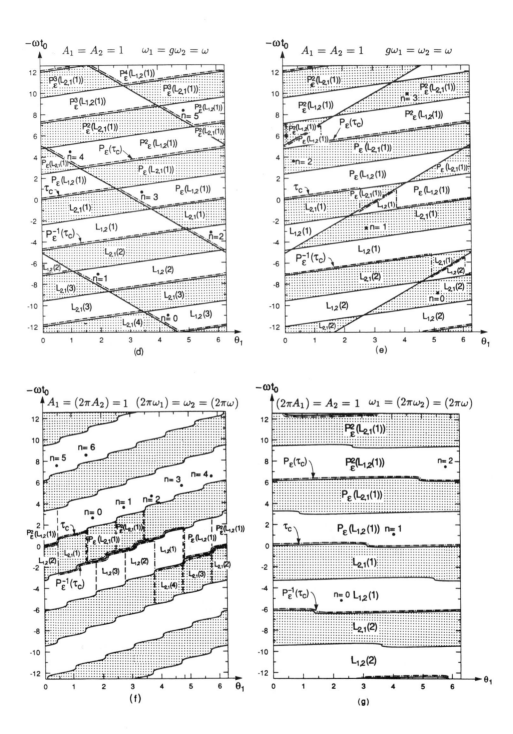

Similarly, the average flux across the surface $\mathcal{S}$, from R_2 into R_1, $\phi_{2,1}(\theta_1)$, is given by

$$\phi_{2,1}(\theta_1) = \langle \theta_{2,1} \left(\theta_1 + 2\pi n \frac{\omega_1}{\omega_2} \right) \rangle_n.$$

These quantities can be computed numerically using the *double phase slice method* described in Beigie *et al.* [1991]. However, using the quasiperiodic Melnikov function, it is shown in Beigie *et al.* [1991] that these quantities can be approximately calculated. Let $L(\theta_1)$ be a two-dimensional lobe on the phase slice $\chi(\theta_1)$ and let $t_0^a(\theta_1)$ and $t_0^b(\theta_1)$ denote the t_0 values of the quasiperiodic Melnikov function corresponding to the pip's that define $L(\theta_1)$, then we have

$$\mu(L(\theta_1)) = \epsilon \int_{t_0^a(\theta_1)}^{t_0^b(\theta_1)} |M(t_0, \theta_1; \theta_{20})| dt_0 + \mathcal{O}(\epsilon^2).$$

Using this result, one can approximate instantaneous flux values. Moreover, we have the following results for the average flux.

Commensurate Frequencies

In this case, under the action of the Poincaré map, the phase slices are mapped only to a finite number of phase slices, say N, and we have

$$\phi_{1,2}(\theta_1) = \phi_{2,1}(\theta_1) = \frac{\omega_2}{2\pi} \frac{\epsilon}{2N} \int_{t_0^0(\theta_1)}^{t_0^{-N}(\theta_1)} |M(t_0, \theta_1; \theta_{20})| dt_0 + \mathcal{O}(\epsilon^2)$$

where

$$t_0^{-N} \text{ is the } t_0 \text{ value of } P_\epsilon^{-N}(\tau_c) \cap \chi(\theta_1).$$

Incommensurate Frequencies

In this case we have

$$\phi_{1,2}(\theta_1) = \phi_{2,1}(\theta_1) = \lim_{T \to \infty} \frac{\epsilon}{2T} \int_0^T |M(t_0, \theta_1; \theta_{20})| dt_0 + \mathcal{O}(\epsilon^2).$$

If all frequencies are mutually incommensurate, then the $\phi_{1,2}$ and $\phi_{2,1}$ are independent of θ_1.

Finally, we remark that we have not made any mention of *chaos* for quasiperiodically time dependent vector fields. There is a generalization of the horseshoe construction that applies; we refer the reader to Beigie *et al.* [1991], Beigie *et al.* [1992], Ide and Wiggins [1989], Meyer and Sell [1989], Scheurle [1986], Stoffer [1988a], [1988b], Wiggins [1988a] and Yagasaki [1992].

CHAPTER 5

Phase Space Transport in the Quasiperiodically Forced Morse Oscillator

5.1. Introduction

In this chapter we apply the methods developed in the previous chapter to a study of the driven Morse oscillator. The Morse potential is a standard model for describing interatomic forces in molecules. The driving terms model an electromagnetic wave (e.g., a laser) interacting with the molecule. More information on this example can be found in Beigie and Wiggins [1992].

Applications of modern dynamical systems theory to problems in theoretical chemistry involving intramolecular energy transfer and the interaction of atoms and molecules with electromagnetic radiation are not new. (See Beigie and Wiggins [1992] and Wiggins [1992] for a more complete bibliography.) However, the vast majority of this work is for time-periodic one-degree-of-freedom systems or two degree-of-freedom systems. In both cases, the problem is reduced to the study of a two-dimensional Poincaré map. For quasiperiodically time-dependent one-degree-of-freedom systems new techniques are needed. Of course, one could ask the question:

Is there any new physical phenomena in multi-frequency time-dependent vector fields as opposed to single frequency vector fields?

We will see that the answer to this question is "yes".

One could also phrase this question in the form of a "thought problem" in the context of the chemistry application discussed in this chapter. One could view the periodically driven Morse oscillator as a model of a diatomic molecule interacting with a laser (single frequency). In this case, one is interested in the transition from bounded motion to unbounded motion. This corresponds to *photodissociation* of the molecule. Now consider an additional, independent laser irradiating the molecule. Can the addition of a second laser, with variable amplitude, frequency and phase, be used to *control* photodissociation?

5.2. The Morse Oscillator

We consider a quasiperiodically forced damped Morse oscillator:

$$\dot{x} = \frac{p}{m},$$

$$\dot{p} = -2Da(e^{-ax} - e^{-2ax}) + \epsilon d(E_1\cos(\omega_1 t + \theta_{10}) + E_2\cos(\omega_2 t + \theta_{20})), \quad (5.1)$$

$(x, p) \in \mathbb{R}^2$. One can think of $x \equiv r - r_e$ as the separation, r, of a two-atom molecule minus an equilibrium separation, r_e, with p the relative momentum. The system then corresponds to a non-rotating pair of atoms interacting under a Morse potential and forced by an external two-frequency electromagnetic field with amplitudes ϵE_1 and ϵE_2. The parameters a and D correspond to the range parameter and *unperturbed* dissociation energy, respectively, of the Morse potential, and d is the effective charge, or dipole gradient. The initial phases associated with the forcing are given by θ_{10}, θ_{20}. For the purpose of numerical calculations, we will use the Morse potential parameters for the HF molecule given by $m = 0.9571$amu, $D = 6.125eV$, $a = 1.1741 r_B^{-1}$ ($r_B \equiv$ Bohr radius), and $d = 0.7876 Dr_B^{-1}$.

The Morse oscillator in autonomous form in the extended phase space is given by:

$$\begin{aligned}
\dot{x} &= \frac{p}{m}, \\
\dot{p} &= -2Da(e^{-ax} - e^{-2ax}) + \epsilon d(E_1\cos\theta_1 + E_2\cos\theta_2), \\
\dot{\theta}_1 &= \omega_1, \\
\dot{\theta}_2 &= \omega_2, \quad (5.2)
\end{aligned}$$

where $(\theta_1, \theta_2) \in T^2$. We then define a Poincaré section:

$$\Sigma^{\theta_{20}} = \{(x, p, \theta_1, \theta_2) \mid \theta_2 = \theta_{20}\},$$

and the associated Poincaré map generated by the trajectories of (5.2) is given by:

$$P_\epsilon : \Sigma^{\theta_{20}} \to \Sigma^{\theta_{20}} \quad (5.3)$$

$$(x(0), p(0), \theta_{10}) \to (x(\frac{2\pi}{\omega_2}), p(\frac{2\pi}{\omega_2}), \theta_{10} + 2\pi\frac{\omega_1}{\omega_2}).$$

Phase Space Structure of the Unperturbed Poincaré Map

The phase portrait of the unperturbed Poincaré map $P_{\epsilon=0}$ is portrayed in Figure 5.1.

There is a neutrally stable one-torus at $\{(x, p, \theta_1) \mid x = p = 0\}$ and a non-hyperbolic invariant one-torus (of saddle-type stability) of the form

$$\mathcal{T}_{\epsilon=0} = \{(x, p, \theta_1) \mid \lim x \to \infty, \ p = 0\}. \quad (5.4)$$

The global stable and unstable manifolds of $\mathcal{T}_{\epsilon=0}$, denoted $W^s(\mathcal{T}_{\epsilon=0})$ and $W^u(\mathcal{T}_{\epsilon=0})$ respectively, coincide to form a two-dimensional separatrix which *separates bounded and unbounded motion*, and which asymptotes with increasing x to $p = 0$.

Any point inside the separatrix evolves on a two-torus and corresponds to a molecule which does not dissociate. Any point outside the separatrix evolves on an

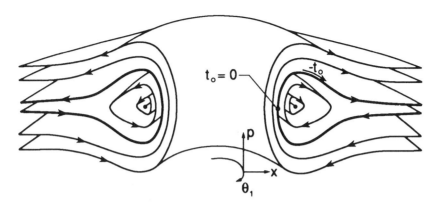

Figure 5.1. Phase space structure for the unperturbed Poincaré map. Also shown is the parametrization for the homoclinic coordinates.

unbounded two-dimensional surface, and corresponds to a molecule that is dissociated, asymptoting to infinite separation. As shown in Figure 5.1, and as discussed earlier, the separatrix is parametrized by (t_0, θ_1) ($t_0 = 0$ is chosen to correspond to the point of symmetry $(x = -\frac{\ln(2)}{a}, p = 0)$).

For the quasiperiodically forced Morse oscillator (5.2) the quasiperiodic Melnikov function is easily calculated and found to be

$$M(t_0, \theta_1, \theta_2; \nu) = -\frac{2\pi}{a}\{E_1 de^{-\omega_1/\omega_0}\sin(\omega_1 t_0 + \theta_1) + E_2 de^{-\omega_2/\omega_0}\sin(\omega_2 t_0 + \theta_2)\} \quad (5.5)$$

where $\omega_0 = a\sqrt{\frac{2D}{m}}$ is the linearized frequency at the bottom of the potential well and the argument ν denotes all the parameters. The trigonometric dependence of the perturbation on $(t, \theta_{10}, \theta_{20})$ in (5.1) carries through to a similar trigonometric dependence of M on $(t_0, \theta_1, \theta_2)$ in (5.5) (the cosines go to sines). We refer to the absolute value of the ratio of the Melnikov amplitude in (5.5), $A_i \equiv -\frac{2\pi}{a}E_i de^{-\omega_i/\omega_0}$, $i = 1, 2$, to the corresponding relative forcing amplitude $E_i d$ in (5.1) or (5.2) as a *relative scaling factor*:

$$relative \ scaling \ factor \ associated \ with \ \omega_i$$

$$\equiv \left| \frac{Melnikov \ amplitude \ for \ \omega_i}{relative \ forcing \ amplitude \ for \ \omega_i} \right| \quad (5.6)$$

$$= \frac{2\pi}{a}e^{-\omega_i/\omega_0}.$$

The dependence of these factors on forcing frequency is described by a single *relative scaling function*:

$$\mathcal{RSF}(\omega; a, \omega_0) = \frac{2\pi}{a}e^{-\omega/\omega_0}. \quad (5.7)$$

The relative scaling function provides an approximate ($\mathcal{O}(\epsilon)$) measure of the effectiveness of a forcing frequency in producing manifold separation, and, as such, is

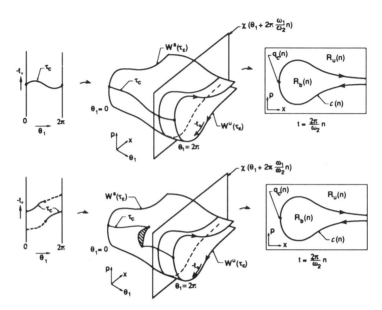

Figure 5.2. An illustration of the division of the Poincaré section, and the phase slices, into regions for a) a toral pim, and b) a spiral pim.

a basic tool in understanding how each frequency affects dissociation. Since the function's exponential decay depends only on ω_0, the period associated with simple harmonic motion at the bottom of the Morse well provides the relevant time scale for the forcing's effectiveness at producing manifold separation. Given any two forcing frequencies, one immediately knows the relative importance of each one; for example, if one of the frequencies is at ω_0 and the other at $4\omega_0$, and the amplitudes of the two forcing terms are identical, then due to the exponential decay of the relative scaling function, the second term has a relatively negligible effect on manifold separation, and hence, as we shall later see, on certain transport properties, so that the problem is essentially one of single frequency forcing. As another example, in the microwave limit $\omega_i \ll \omega_0$ the relative scaling factor associated with each frequency will be approximately $\frac{2\pi}{a}$, essentially frequency independent.

5.3.   Set-Up of the Transport Problem

Here we set up the partial barrier between regions of phase space corresponding to bounded and unbounded motion and subsequently construct the turnstiles controlling access to these regions. Once this is done, transport and flux between these regions can be discussed.

We let $S[\tau_\epsilon, \tau_c]$ and $U[\tau_\epsilon, \tau_c]$ denote the segments of $W^s(\tau_\epsilon)$ and $W^u(\tau_\epsilon)$, respectively, from τ_ϵ to τ_c, then $\mathcal{C} \equiv U[\tau_\epsilon, \tau_c] \cup S[\tau_\epsilon, \tau_c]$ denotes a two dimensional surface in $\Sigma^{\theta_{20}}$ that divides *each phase slice* $\chi(\bar{\theta}_1)$, $\bar{\theta}_1 \in T$, into two regions, as illustrated in Figure 5.2.

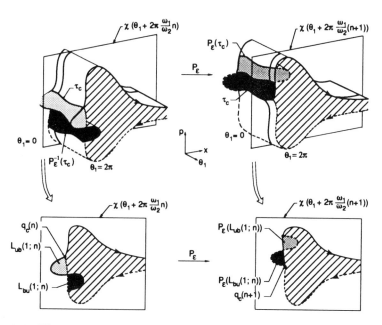

Figure 5.3. Illustration of the turnstile construction, and their dynamics, for the case of a toral pim.

For the case of toral pim's, illustrated in Figure 5.2(a), one can choose τ_c to be a one-torus, and $\mathcal{C}$ in fact divides $\Sigma^{\theta_{20}}$ into two regions; for the case of spiral pim's, illustrated in Figure 5.2(b), one must choose τ_c to have a discontinuity, and hence $\mathcal{C}$ is discontinuous and does *not* divide $\Sigma^{\theta_{20}}$ into two regions, since there are gaps at the region of discontinuity. However, for all cases, $\mathcal{C}$ divides each *phase slice* $\chi(\bar{\theta}_1)$, $\bar{\theta}_1 \in T$, into two regions. At each instant of discrete time $t = \frac{2\pi}{\omega_2} n$, we thus have in (x, p) space a time dependent boundary $\mathcal{C}(n) \equiv \mathcal{C} \cap \chi(\theta_1 + 2\pi \frac{\omega_1}{\omega_2} n)$ that divides the space into two regions, denoted $R_b(n)$ and $R_u(n)$, as illustrated in Figure 5.2. (Note that invariant three-dimensional regions in $\Sigma^{\theta_{20}}$, R_u and R_b, are then defined as the union of the corresponding two-dimensional regions over the phase slices defined by $\bar{\theta}_1 \in T$.)

As explained in the last chapter, points can cross $\mathcal{C}$ *only* via the *turnstile lobes*, which we now define and identify. We will adopt a notation that is slightly different from that given in the previous chapter; however, it has the advantage of being less cumbersome and more natural for the specific problem under consideration. It is a straightforward consequence of the orientation-preserving nature of P_ϵ that the lobes between τ_c and $P_\epsilon^{-1}(\tau_c)$, which we refer to as the *turnstiles*, change their orientation *relative to* $\mathcal{C}$ under P_ϵ. What we mean by this can be easily visualized in Figure 5.3, where we show an example with toral pim's and one pair of turnstile lobes. It should be clear from the figure that points in the turnstiles, and *only* these points, cross the invariant boundary $\mathcal{C}$ under P_ϵ, mapping from outside to inside or vice-versa. At each instant of discrete time $t = \frac{2\pi}{\omega_2} n$, we thus have in (x, p) space time-dependent turnstiles defined by

$$L_{bu}(1;n) \equiv L_{bu}(1) \cap \chi(\theta_1 + 2\pi \frac{\omega_1}{\omega_2} n),$$

$$L_{ub}(1;n) \equiv L_{ub}(1) \cap \chi(\theta_1 + 2\pi \frac{\omega_1}{\omega_2} n), \tag{5.8}$$

where $L_{bu}(1)$ and $L_{ub}(1)$ are the three-dimensional turnstile lobes in $\Sigma^{\theta_{20}}$ that map under one iterate of P_ϵ from R_b to R_u and from R_u to R_b, respectively. Hence the turnstile $L_{bu}(1;n)$ ($L_{ub}(1;n)$) is the set of lobes which map from *inside* (*outside*) $R_b(n)$ to *outside* (*inside*) $R_b(n+1)$, and we shall refer to these two processes as *escape* and *capture*, respectively. The points in $R_b(n)$ at time $t = \frac{2\pi}{\omega_2} n$ are destined to oscillate in a bounded fashion until at some future time $t = \frac{2\pi}{\omega_2} \tilde{n}$, $\tilde{n} > n$ (which may or may not ever occur), they *escape via the turnstile lobes* $L_{bu}(1; \tilde{n})$ to $R_u(\tilde{n}+1)$ and henceforth asymptote to infinite separation. (In a similar manner one describes capture.) Hence, it is correct to interpret $R_b(n)$, $R_u(n)$ as the regions of bounded and unbounded motion, respectively, and the turnstile lobes $L_{bu}(1;n)$ and $L_{ub}(1;n)$ as the *sole* mechanism for transport between the bounded and unbounded regions.

5.4.   Fluxes and Dissociation rates

We will consider flux of phase space from R_u into R_b. In the case of quasiperiodically time-dependent vector fields there are five relevant, and in general *different*, measures of flux. The first two are the *instantaneous flux*, denoted $\phi_e(n)$ and $\phi_c(n)$ for escape and capture, respectively:

$$\phi_e(n) = \frac{\omega_2}{2\pi} \mu(L_{bu}(1;n)),$$

$$\phi_c(n) = \frac{\omega_2}{2\pi} \mu(L_{ub}(1;n)), \tag{5.9}$$

where $\mu(\cdot)$ denotes the area of the lobes within the parenthesis. Instantaneous flux thus refers to the volume in phase space, per discrete time unit, that escapes or is captured between the nth and $(n+1)^{th}$ discrete time samples, and is in general different for each time sample and for escape versus capture, since the partial barriers vary with each time slice and the area of the region they enclose can change. The next two measures of flux are the *finite-time average flux*, denoted $\Phi_e(n)$ and $\Phi_c(n)$ for escape and capture, respectively, which are simply the average of the instantaneous flux over the first n time samples, i.e.,

$$\Phi_e(n) = \frac{1}{n} \sum_{i=1}^n \phi_e(i),$$

$$\Phi_c(n) = \frac{1}{n} \sum_{i=1}^n \phi_c(i). \tag{5.10}$$

These, as well, are different for each n and for escape versus capture, and though both quantities will converge to the same value in the limit $n \to \infty$, the convergence time can vary from being quite short to quite long. The final measure is the *infinite-time average flux*

$$\Phi = \lim_{N \to \infty} \frac{1}{2N} \sum_{n=0}^{N} \{\phi_e(n) + \phi_c(n)\}, \tag{5.11}$$

which is the same for escape and capture, since the area of $R_b(n)$ remains bounded, as should be plain from the definition of $R_b(n)$. One can exactly compute the above fluxes by identifying the turnstiles and computing their boundaries for each phase slice using the numerical *double phase slice method* as described in Beigie *et al.* [1991].

We use this method to compute the different types of fluxes numerically and make a comparison between transport in one versus two frequency vector fields.

Infinite-Time Average Flux–One vs. Two Frequencies

We want to compare the infinite-time average flux associated with periodic and quasiperiodic forcing. A normalization criterion for the forcing amplitudes is needed to decide upon "equivalent" periodic and quasiperiodic perturbations that can be compared. For example, let us choose the criterion $E_1^2 + E_2^2 = constant \equiv E^2$ (i.e. as E_1 and E_2 are varied, we keep constant the sum of the intensities associated with each amplitude in the electromagnetic field). For a fixed ω_1, ω_2, we then vary, say, E_1 from 0 to E, with E_2 determined by the normalization criterion. The question then is: How does the infinite-time average flux vary as a function of E_1? The variation depends on two properties: interference effects and relative scaling factors. Referring to Figure 5.4, at the single frequency limits, associated with the endpoints $E_1 = 0$ and $E_1 = E$, the average flux Φ is proportional to the relative scaling factor associated with the corresponding frequency. Interference effects cause the flux profile to dip below a linear increase from the lower to higher Φ. (This holds for all ω_1, ω_2, $\theta_{10} - \theta_{20}$, and for all $E_1 \in (0, E)$.) The single frequency limit associated with the larger relative scaling factor thus corresponds to an absolute maximum of the average flux. (The absolute minimum may or may not be at the other single frequency limit, depending on the difference between the two relative scaling factors and the size of the dip due to interference effects.) In this context, the best one can do is single frequency forcing.

Finite-Time Average Flux–Control of Escape vs. Capture

Since the finite-time escape flux need not equal the finite-time capture flux, the possibility arises that the ratio of these two quantities may be controlled by appropriate choice of parameters. In Figure 5.5 we show that this is the case.

Penetration of the Lobes into the Potential Well

The Melnikov function (in all its various settings) is a measure of the splitting of the perturbed stable and unstable manifolds in the direction normal to the unperturbed separatrix. Hence, in the context of this problem, it is a measure of

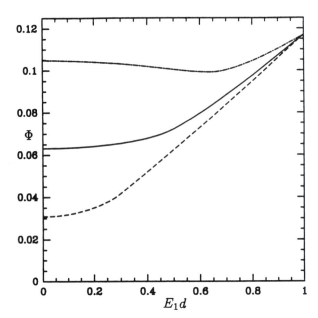

Figure 5.4. Infinite-time average flux as a function of $E_1 d$, with $(E_1^2 + E_2^2)d^2 \equiv E^2 d^2$. The frequency ω_1 is fixed at ω_0; the frequency ω_2 is $\frac{7}{3}\omega_0$ (dashed), ω_0/g (solid), and $\frac{10}{9}\omega_0$ (dashed-dotted). The flux is per unit $\epsilon \frac{2\pi}{a} E d$, and $\theta_{10} = \theta_{20} = 0$.

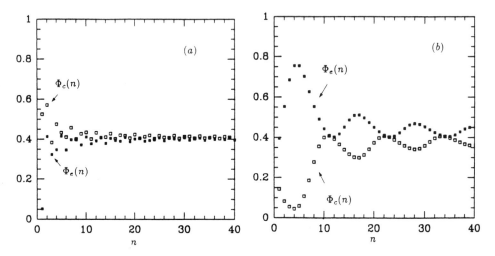

Figure 5.5. The variation of convergence rate for finite-time average flux. The black dots correspond to escape, the white dots to capture, and the flux is per unit ϵA. For this example $(A_1, A_2) \equiv -(A, A)$, $A \equiv -\frac{2\pi}{a} E d e^{-\omega/\omega_0}$, and (a) $(\omega_1, \omega_2) = (g, 1)\omega_0$, $(\theta_{10}, \theta_{20}) = (\pi, 0)$, (b) $(\omega_1, \omega_2) = (0.231, 2.618)\omega_0$, and $(\theta_{10}, \theta_{20}) = (\frac{23}{22}\pi, \frac{3}{2}\pi)$.

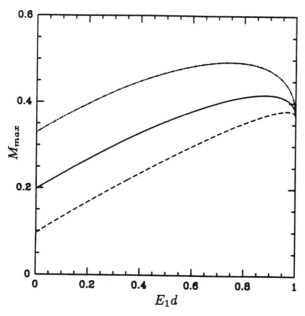

Figure 5.6. Plot of $M_{\max}$ as a function of $E_1 d$, with $(E_1^2 + E_2^2)d^2 \equiv E^2 d^2$. The parameters are identical to those in the plot of the infinite-time average flux. $M_{\max}$ is per unit $\frac{2\pi}{a} Ed$.

how far the perturbed manifolds penetrate into the potential well. From this, it follows that it is a measure of the depth of the penetration of the turnstiles into the potential well. If this depth of penetration into the well can be controlled by additional frequencies, then photodissociation of particular unperturbed energy levels can similarly be controlled. In Figure 5.6 we plot the maximum of the quasiperiodic Melnikov function, $M_{\max}$, for different parameters values. From this we see that the maximum occurs in the case of two-frequency forcing. Hence lower energy levels can photodissociate in the two-frequency case. We remark that our brief discussion gives the general idea, but does not touch on the many subtleties of this problem of penetration of the manifolds into the potential well. There are also some important time-scale considerations. For more details see Beigie and Wiggins [1992].

Summary Comparison of Single-Frequency vs. Two-Frequency Transport

- Infinite-time average flux is largest in the single-frequency limit corresponding to the largest relative scaling factor (provided the amplitude of the forcing is normalized as described above).

- The ratio of "capture" vs. "escape" flux can be varied by appropriate parameter choices *on finite time scales.*

- Two-frequency forcing can result in a deeper penetration of the manifolds into the potential well, as opposed to single-frequency forcing.

CHAPTER 6

Adiabatic Dynamical Systems

Here we consider a very different type of vector field than those that we have considered previously. Consider a two-dimensional, slow time-varying Hamiltonian system with a Hamiltonian of the form

$$H(x, y, \epsilon t).$$

When t varies by an $\mathcal{O}(\frac{1}{\epsilon})$ amount the Hamiltonian may vary by an $\mathcal{O}(1)$ amount. Hence the "perturbation" is large, but this large variation occurs slowly. This is in contrast to the more familiar Melnikov methods where the maximum variation in the perturbation amplitude is $\mathcal{O}(\epsilon)$.

In our earlier treatment of time-dependent vector fields we *suspended* the system by introducing the phases of the different frequency components of the time-dependence as new *dependent* variables. This resulted in an enlarged phase space, but from the point of view of geometry it allowed us to treat the unperturbed and perturbed systems on a more equal footing. For these *adiabatic dynamical systems* the suspension is of a very different character, and it is responsible for us being able to treat "large perturbations".

We suspend the system by introducing the new dependent variable $z = \epsilon t$. Hamilton's equations are then given by

$$\dot{x} = \frac{\partial H}{\partial y}(x, y, z),$$

$$\dot{y} = -\frac{\partial H}{\partial x}(x, y, z), \qquad (6.1)$$

$$\dot{z} = \epsilon,$$

where we will assume $z \in T$ (the one-torus) and $(x, y) \subset K$, where K is some compact set in $\mathbb{R}^2$. We remark that we can treat situations where z is not an angular variable; we will give references for these at the end of this chapter. However, the periodic case will suffice for introducing the main ideas.

The Unperturbed Structure

We make the following assumption on the unperturbed system, i.e., (6.1) with $\epsilon = 0$.

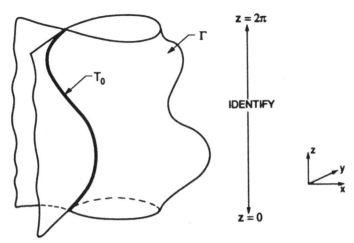

Figure 6.1. The parametrization of Γ for an adiabatically time-varying vector field.

Assumption. At $\epsilon = 0$ (6.1) has a hyperbolic fixed point for each $z \in T$, denoted $(x(z), y(z))$, which is connected to itself by a homoclinic orbit $(x^h(t; z), y^h(t; z))$. We also assume that $(x(z), y(z)) \subset K$ for each $z \in T$.

In this case,

$$T_0 = \{(x, y, z) \,|\, x = x(z),\ y = y(z),\ z \in T\}$$

is a normally hyperbolic invariant one-torus (circle) of fixed points having two-dimensional stable and unstable manifolds denoted $W^s(T_0)$, $W^u(T_0)$, respectively, that coincide along a branch forming a two-dimensional homoclinic manifold that can be parametrically represented using the unperturbed homoclinic orbits as follows:

$$\Gamma = \left\{(x, y, z) \,|\, x = x^h(-t_0; z),\ y = y^h(-t_0; z),\ t_0 \in \mathbb{R},\ z \in T\right\}.$$

The parametrization is illustrated in Figure 6.1.

Now we point out an important feature of these adiabatic dynamical systems. Our unperturbed structure "breaths" or changes as z (the variable that "enlarges" the phase space) is varied. (This idea should be intuitively clear from Figure 6.1.) This is very different from the situation for time-periodic or quasiperiodic vector fields where the unperturbed phase space structure was identical for all values of the phases of the different frequency components (the variables that "enlarge" the phase space). Since z is like time, our phase space structure about which we develop our perturbation methods varies "statically" in such a way that it is "close" ($\mathcal{O}(\epsilon)$) to the perturbed structure. The net result of this is that, in the perturbed problem, the manifolds can vary, in relation to the unperturbed manifolds for fixed z, over an $\mathcal{O}(1)$ region of the phase space.

The Adiabatic Melnikov Function

One can develop a measure of the splitting of the stable and unstable manifolds of the perturbed one-torus in exactly the same way as described earlier for time-periodically and quasiperiodically perturbed vector fields. This is done in the following steps:

1. Use the unperturbed homoclinic orbits to develop "homoclinic coordinates" for describing the perturbed structure.

2. Appeal to the persistence theory for normally hyperbolic invariant manifolds, and transversal intersection of manifolds on compact sets.

3. Use 1) and 2) together to develop a measure of the distance between the perturbed stable and unstable manifolds.

4. Taylor expand this distance in the perturbation parameter and use "Melnikov's trick" to develop a *computable* form for the leading order term in this distance measurement.

For the details of this procedure, we refer the reader to Wiggins [1988b]. Here, we merely give the result.

The Melnikov function that measures the distance between the stable and unstable manifolds of the perturbed hyperbolic orbit is given by

$$M_A(z) = \int_{-\infty}^{+\infty} t \left\{ H, \frac{\partial H}{\partial z} \right\} (x^h(t;z), y^h(t;z), z) dt, \tag{6.2}$$

where $\left\{ H, \frac{\partial H}{\partial z} \right\} \equiv \frac{\partial H}{\partial x} \frac{\partial^2 H}{\partial y \partial z} - \frac{\partial H}{\partial y} \frac{\partial^2 H}{\partial x \partial z}$ is the *Poisson bracket*.

An Example

We give an example of a pendulum with a slowly varying base support which illustrates some further unique aspects of adiabatic dynamical systems. This system is a Hamiltonian system with Hamiltonian $H(x, y, ; z) = \frac{y^2}{2} - (1 - \gamma \cos z) \cos x$ and described by the following equations:

$$
\begin{aligned}
\dot{x} &= y, \\
\dot{y} &= -(1 - \gamma \cos z) \sin x, \\
\dot{z} &= \epsilon,
\end{aligned}
\tag{6.3}
$$

where $0 < \gamma < 1$. At $\epsilon = 0$ (6.3) has a hyperbolic fixed point at $(\pi, 0) = (-\pi, 0)$ for all values of z. Moreover, these fixed points are connected together by the homoclinic orbits:

$$
\begin{aligned}
x^h(t;z) &= \pm 2\sqrt{1 - \gamma \cos z} \operatorname{sech}(1 - \gamma \cos z \, t), \\
y^h(t;z) &= \pm 2 \arcsin\left(\tanh(1 - \gamma \cos z \, t)\right).
\end{aligned}
$$

Using these expressions along with (6.2), the adiabatic Melnikov function is easily calculated, and is given by

$$M_A(z) = \frac{4\gamma \sin z}{\sqrt{1 - \gamma \cos z}}. \tag{6.4}$$

Clearly, $M_A(z)$ has simple zeros with respect to z and therefore (6.3) has transverse homoclinic orbits to a one-torus (periodic orbit). Hence, there exists chaos in the sense of Smale horseshoes. Since the chaos occurs over such long time scales ($\mathcal{O}(\frac{1}{\epsilon})$) the chaos in this type of system has been referred to as *adiabatic chaos* (Wiggins [1988c]).

Lobe Dynamics and Turnstiles

The vector fields that we are considering are time-periodically perturbed one-degree-of-freedom Hamiltonian systems–just as we considered in Chapter 2. Thus, they can be reduced to the study of a two-dimensional Poincaré map. Therefore the lobe turnstile, and transport formulae in terms of turnstile dynamics discussed in Chapter 2 should all go through here, and for the most part it does, except for a few wrinkles due to the $\mathcal{O}(\epsilon)$ frequency. Let us first look more closely at the Poincaré map.

The associated Poincaré map of (6.3) is given by

$$(x(0), y(0)) \mapsto \left(x\left(\frac{2\pi}{\epsilon}\right), y\left(\frac{2\pi}{\epsilon}\right) \right).$$

Now for ϵ small, the return time for this Poincaré map is very large. In fact, the return time goes to ∞ as the perturbation strength goes to zero. This fact allows for the lobes associated with the homoclinic tangle to become drastically stretched and folded. Since the interior of the region bounded by the homoclinic orbits is of finite size, and KAM tori exist fairly close to the separatrix (since the frequency of the forcing is ϵ), the outgoing portion of the turnstile has little choice but to wrap itself throughout the interior of the ingoing portion of the turnstile. This is an example of what we referred to as a *self-intersecting turnstile* in Chapter 2, and we will see how it can be dealt with in a simple way in the context of an example in the next chapter. We demonstrate this phenomenon numerically for the slowly varying pendulum in Figure 6.2 for parameter values $\gamma = 0.75, \omega = 2\pi$ and $\epsilon = \frac{1}{12}$.

We expect this phenomenon of the self-intersection of turnstiles to typically occur in "periodically adiabatically forced" systems such as (6.3). The reason being that since the frequency of forcing is small the return time of the associated Poincaré map is large. This in turn may result in considerable stretching and folding of the lobes between iterations.

Some Comments on Adiabatic Invariants

In a region of the phase space inside and bounded away from the separatrices a z-dependent canonical transformation can be made that transforms (6.1), at $\epsilon = 0$,

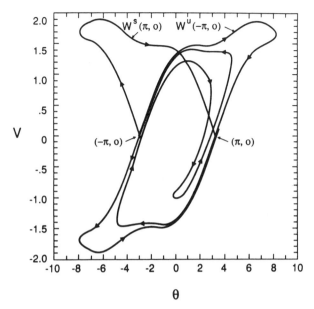

Figure 6.2. Self-intersecting turnstile for the parametrically, adiabatically forced pendulum (computation by T. Kaper and D. Hobson).

to the following form:

$$
\begin{aligned}
\dot{I} &= 0, \\
\dot{\theta} &= \Omega(I, z), \\
\dot{z} &= 0.
\end{aligned}
\tag{6.5}
$$

In other words, we have a z-dependent transformation to action-angle variables that follows from the integrable structure. Each I value labels an invariant two-torus in the region of phase space in which the $I - \theta - z$ variables are valid, for which we must have $\Omega(I, z)$ bounded away from zero. Arnold [1962] developed a modification of the proof of the KAM theorem which showed that for ϵ small many of these tori are preserved (i.e., those which have sufficiently irrational rotation number, which form a majority in the sense of Lebesgue measure). Since these *Arnold tori* are co-dimension 1, they form boundaries which trajectories cannot cross and therefore from Arnold's theorem one can conclude that the action is an adiabatic invariant *for all time*. We stress that the preservation of tori in this setting is different than the standard KAM setting as in the unperturbed problem all of the tori are resonant ($\dot{z} = 0$).

Final Remarks

1. A selection of references on adiabatic dynamical systems is as follows: Bruhwiler and Cary [1989], Cary *et al.* [1986], Cary and Skodje [1989], Elskens and

Escande [1991], Kaper and Wiggins [1992], [1993], Neishtadt [1975], Neishtadt *et al.* [1991], Palmer [1986], Robinson [1983], Wiggins [1988a], [1988b], [1988c].

2. Kaper and Wiggins [1992] have shown that the integral of the adiabatic Melnikov function between two adjacent zeros (an $\mathcal{O}(1)$ quantity) is an approximation to the lobe area for adiabatic dynamical systems. We will see a use for this result in the next chapter.

3. Upon examining the references in 1) and 2) one sees that a useful analytical framework for the study of adiabatic dynamical systems is that of *geometric singular perturbation theory*, or, in other words, the study of foliations of normally hyperbolic invariant manifolds. These techniques allow one to turn singular perturbation problems into regular perturbation problems. (See Fenichel [1979], Wiggins [1993], and references therein.)

CHAPTER 7

Transport in the Eccentric Journal Bearing Flow: An Application of Adiabatic Dynamical Systems

7.1. Introduction

In this chapter we describe transport in an adiabatic dynamical system in the context of a fluid mechanical example, much as in Chapter 2. The system we study is the *eccentric journal bearing flow*. The system consists of two cylinders, one inside the other, and the term *eccentric* comes from the fact that the axes of rotation of the cylinders are not identical. We refer to the outer cylinder as the *casing* and the inner cylinder as the *shaft*. We are interested in studying the motion of fluid particles in between the cylinders. We will consider a special type of fluid flow, a *Stokes flow*. These can be thought of as highly viscous flows characterized by the magnitudes of certain dimensionless quantities (low Reynolds number). In this parameter regime, certain terms in the Navier-Stokes equation can be neglected (the Stokes flow approximation) and the resulting equations are highly amenable to solution by asymptotic methods (e.g. Leal [1992]). In particular, streamfunctions for the eccentric journal bearing flow have been obtained by Ballal and Rivlin [1976] for the case of cylinders rotating at constant angular velocities. In this case the flow is two-dimensional and steady (i.e., the streamfunction is independent of time). If the rotation rates vary sufficiently *slowly* in time, then it can be shown that the Stokes approximation remains valid. We can then use the streamfunctions obtained by Ballal and Rivlin [1976], but modified by allowing the frequencies to vary in time. We refer to a specific way in which the rotation rates vary in time as a *modulation protocol*. More discussion, and references, can be found in Kaper and Wiggins [1993] and the discussion in this chapter is based on this work.

7.2. The Eccentric Journal Bearing Flow

The Steady Counterrotating Flow

The integrable equations for the evolution of fluid particles in the steady flow are:

$$\dot{x} \;=\; \frac{\partial \psi}{\partial y}(x, y; \Omega_1, \Omega_2),$$

$$\dot{y} = -\frac{\partial\psi}{\partial x}(x, y; \Omega_1, \Omega_2) \tag{7.1}$$

where ψ is the streamfunction, x and y are Cartesian coordinates, and Ω_1 and Ω_2 are the angular velocities of the casing and the shaft, respectively.

The Reynolds number is defined as:

$$\text{Re} \equiv \frac{\left[R_1^2\Omega_1^2 + R_2^2\Omega_2^2\right]^{\frac{1}{2}} R_2}{\nu},$$

where R_1 and R_2 denote the radii of the casing and shaft, respectively, with $R_1 > R_2$, and ν is the kinematic viscosity. In using the results of Ballal and Rivlin [1976], we take $0 < \text{Re} \ll 1$, which is the limit of steady Stokes flow.

The parameters for the steady flow are: $\bar{e} = \frac{\Delta x}{R_1 - R_2}$, the eccentricity of the bearing, where Δx is the distance (measured in the same cartesian coordinates shown in the figures) between the centers of the cylinders; $\bar{\Omega} \equiv \frac{\Omega_2}{\Omega_1}$, the ratio of the angular velocities; and, $\bar{r} \equiv \frac{R_2}{R_1}$, the ratio of the radii of the cylinders. Our $\bar{e}$ and Δx are identical to the variable $\bar{\epsilon}$ and ϵ in Ballal and Rivlin [1976]. Furthermore, for simplicity, the results we report in the present work are for the geometry with $\bar{r} = 0.3$. Simulations using other values of $\bar{r}$ are qualitatively similar to those obtained with the choice $\bar{r} = 0.3$.

In the case of counterrotating shafts ($\Omega_1 \cdot \Omega_2 < 0$) there is precisely one saddle stagnation point $\mathbf{X_0}$ (hyperbolic fixed point) on the x-axis in the narrow gap for all values of $\bar{e}$, $\bar{r}$, Ω_1 and Ω_2, which is attached to itself by two stagnation streamlines (orbits homoclinic to the hyperbolic fixed point). (See Figure 7.1.) These two stagnation streamlines, which are the inner and outer stagnation streamlines Γ and Λ, respectively, separate the fluid domain into three regions: the annular area adjacent to the shaft; the kidney-shaped backflow region; and the annular region adjacent to the casing.

A more complete catalog of formulae and streamline plots, including ones for the corotating case, may be found in Ballal and Rivlin [1976]. The reader may also find the relevant quantities in Appendix A of Part II of Kaper [1991].

The Unsteady Flow: Modulation Protocols

We present results using two modulation protocols in this work:

$$\Omega_1(\epsilon t) = 1, \qquad \Omega_2(\epsilon t) = -6 + 4\cos(\epsilon t), \qquad \text{(MP1)}$$

$$\Omega_1(\epsilon t) = 1, \qquad \Omega_2(\epsilon t) = -30.5 + 29.5\cos(\epsilon t), \qquad \text{(MP2)}$$

where $0 < \epsilon \ll 1$. With the choice of (MP1) and (MP2), we operate well within the range of angular velocities for which Ballal and Rivlin [1976] report their results. Also, quasisteadiness is maintained when the Stokes number $Sto = \frac{\epsilon L^2}{\nu}$ is small.

Introduction of the modulation protocols makes the equations (7.1) noninte-grable and puts them in the form of an adiabatic dynamical system:

$$\dot{x} = \frac{\partial\psi}{\partial y}(x, y; \Omega_1(z), \Omega_2(z)),$$

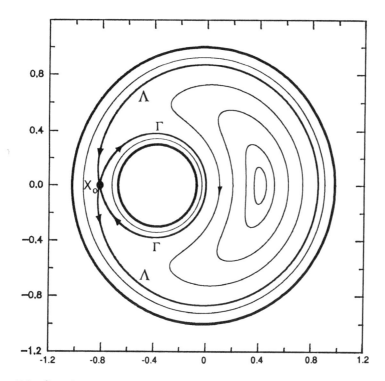

Figure 7.1. Steady state counterrotating eccentric journal bearing flow, with $\bar{e} = 0.5$, $R_1 = 1.0$, $R_2 = 0.3$, $\bar{r} = 0.3$, (where for Ballal and Rivlin $\xi_1 = -0.9397$, $\xi_2 = -1.9966$). The cylinders are the darkest circles. The inner circle rotates clockwise with $\Omega_2 = -4$, and the outer circle rotates counterclockwise with $\Omega_1 = 1$.

$$\dot{y} = -\frac{\partial \psi}{\partial x}(x, y; \Omega_1(z), \Omega_2(z)),$$

$$\dot{z} = \epsilon, \tag{7.2}$$

where the dependence of ψ on z is through the slowly-varying time-dependent functions $\Omega_1(z)$ and $\Omega_2(z)$, given by the modulation protocols. Since the modulation protocols are periodic in $z = \epsilon t$ and (7.2) is Hamiltonian, we use the area-preserving Poincaré map

$$T_{z_0}\begin{pmatrix} x(z_0) \\ y(z_0) \end{pmatrix} \equiv \begin{pmatrix} x(z_0 + 2\pi) \\ y(z_0 + 2\pi) \end{pmatrix}$$

with $z_0 \in [0, 2\pi)$, which gives a stroboscopic picture of the fluid domain.

The flow parameter introduced by these modulations is

$$\Delta\bar{\Omega} \equiv \max_{z \in [0,2\pi)} \Omega_2(z) - \min_{z \in [0,2\pi)} \Omega_2(z).$$

It measures the amplitude of the modulation, since the ratio of Ω_2 to Ω_1 determines the location of the instantaneous saddle. (See Figure 7.2 for some examples.) For (MP1), $\Delta\bar{\Omega} = 8$ because $\bar{\Omega}(z)$ varies in the interval $[-2, -10]$, and $\Delta\bar{\Omega} = 59$ for (MP2) because $\bar{\Omega}(z)$ varies in the interval $[-1, -60]$. Hence, (MP1) is a "moderate" modulation, because the instantaneous saddle stagnation point moves slowly and periodically across a moderately large fraction of the gap between the cylinders, and (MP2) is a "strong" modulation, because the instantaneous saddle stagnation point moves slowly and periodically across almost the entire gap between the cylinders.

We conclude this section by discussing a concept used often in the remainder of the present work, that of an *instantaneous streamline*. The instantaneous streamline is the closed trajectory the particle would execute if the system evolves with the value of z frozen at its instantaneous value. Thus, it coincides with the orbit of the steady state flow (with Ω_1 and Ω_2 equal to their value at the instantaneous value of z) that passes through the particle's instantaneous position. Of course, since particle paths in the modulated flow no longer coincide with streamlines, this concept is a fictitious one; nevertheless it is often used in adiabatic dynamical systems, (Kaper and Wiggins [1992] and Escande and Elskens [1991]), and it is helpful in the context of quasisteady Stokes flows, as well. The inner and outer instantaneous stagnation stream lines at the instant of slow time z are denoted Γ^z and Λ^z, respectively. The analogous concept exists for *instantaneous stagnation points*.

The Potential Mixing Zone

In much of the $\mathcal{O}(1)$ area swept out by the (fictitious) instantaneous streamlines, the theory of adiabatic invariance and Arnold's extension of the KAM theorem to adiabatic Hamiltonian systems are not applicable because one of the main assumptions that those theories rely on ceases to be valid on and near stagnation streamlines, which are zero frequency orbits. The violated assumption is that the frequency of the instantaneous steady state orbits be one order of magnitude larger, i.e., $\mathcal{O}(1)$, than the modulation frequency ϵ. In fact, observations collected from theoretical and numerical work on various model problems from mechanics (Cary *et al.* [1986], Cary and Skodje [1989], Bruhwiler and Cary [1989], Kaper and Wiggins [1992], and Elskens and Escande [1991]) suggest that tracer particles can explore most of this swept out area because there are very few barriers to their transport in this region. Hence this is the region in which mixing can be expected to occur. We caution, however, that there is a piece of the swept out area in which adiabatic invariance does apply, despite the fact that there are instantaneous separatrices there. This is dealt with in Kaper and Wiggins [1993], in which there is also a more detailed discussion of the structure of the potential mixing zone.

For (MP1) and (MP2), the region swept out may be specified exactly as

$$\left(\cup_{z\in[0,\pi]}\Gamma^z\right) \cup \left(\cup_{z\in[0,\pi]}\Lambda^z\right), \tag{7.3}$$

where we recall that Γ^z and Λ^z are the inner and outer instantaneous stagnation streamlines at time z. Because the measure of the set (7.3) is to leading order the area potentially available for mixing, we label the set given by (7.3) as the *potential mixing zone*.

The symmetry in (MP1) and (MP2) about $z = \pi$, namely $\bar{\Omega}(z) = \bar{\Omega}(2\pi - z)$ for $z \in (\pi, 2\pi]$, implies that we only need to take the unions over $z \in [0, \pi]$ in (7.3). For more general protocols, the unions are taken over all z between the two values of z corresponding to an instantaneous separatrix which locally (*i.e.*, compared to the instantaneous separatrices for nearby values of z) encloses a maximum area and an instantaneous separatrix which locally encloses a minimum area (Kaper and Wiggins [1992]). We will see how the potential mixing zone relates to lobe area shortly. In Figure 7.2 we show the potential mixing zone for three different sets of parameter values.

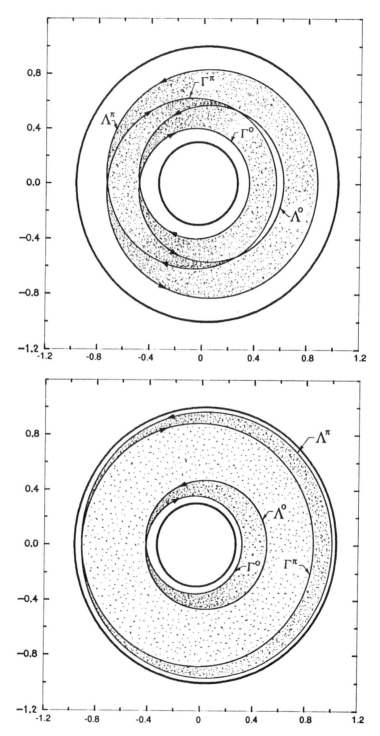

Figure 7.2. Potential mixing zones (all with $\bar{r} = 0.3$). a) $\bar{e} = 0.1$ with (MP1); b) $\bar{e} = 0.1$ with (MP2); c) $\bar{e} = 0.75$ with (MP1). Note that in c) Γ^0 and Λ^π nearly coincide with the inner and outer cylinders, respectively.

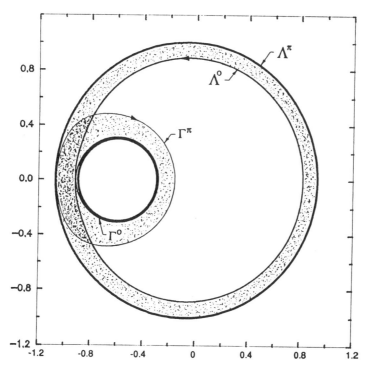

7.3.   Transport Analysis–Lobe Geometry and Dynamics

As we discussed in the previous chapter, the homoclinic tangles one observes in slowly-modulated systems are, both qualitatively and quantitatively, very different from those studied in weakly-perturbed systems. The differences between the tangles in these two types of systems are especially striking in the measurements of the stretching and transport quantities which we give in these sections. Since this structure has only recently been understood (Kaper and Wiggins [1992], and Elskens and Escande [1991]), we describe those parts of the theory we need in this section.

Chief among the differences is that the lobes are very long and thin – in particular what we will define as the turnstile lobes are of length $\mathcal{O}(\frac{1}{\epsilon})$ and area $\mathcal{O}(1)$ so that their average width is $\mathcal{O}(\epsilon)$ asymptotically as $\epsilon \to 0$, as shown in Kaper and Wiggins [1993], and each successive image (both forward and backward) of a turnstile lobe is longer and thinner than the preceding image, while having the same area due to the area-preservation property of the Poincaré map.

In addition, the singular-perturbation nature of these modulated flows implies that the lobe area is the inter-regional flux per half period, and, as a result, the transport in these flows is more complicated than in the usual case in which transport is studied every full period only.

The last difference we mention is that the turnstiles permeate the entire mixing zone, causing the tangle to function as its "backbone," because the tangle sweeps out the entire $\mathcal{O}(1)$ separatrix-swept area. This unique feature of adiabatic dynamical systems guarantees that large scale mixing occurs within the first few periods of the modulation.

Structures in Adiabatic Dynamical Systems

The saddle stagnation point of the steady flow with $z = z_0$ persists as the saddle stagnation point $X_\epsilon(z_0)$ of the Poincaré map T_{z_0}, (*i.e.*, as a periodic trajectory of the flow). See Wiggins [1988a,b] for an exposition of the theory specific to adiabatic dynamical systems. Since $X_\epsilon(z_0)$ lies near $X_0(z_0)$, one can construct an asymptotic expansion in powers of ϵ for its position as a function of z. (See Appendix B of Part II of Kaper [1991].) The leading order term is $X_0(z_0)$, and the first correction term is $\mathcal{O}(\epsilon)$ in the y-component and $\mathcal{O}(\epsilon^2)$ in the x-component. Furthermore, from symmetry considerations, $X_\epsilon(z_0)$ lies on the x–axis for $z_0 = 0 \bmod 2\pi$ and for $z_0 = \pi \bmod 2\pi$.

If one were to watch the experiment continuously as z increases from z_0 to $z_0 + 2\pi$, instead of sampling it stroboscopically with the Poincaré map, one would see that the saddle executes a periodic orbit $\gamma_\epsilon(z)$. This closed path lies in the fluid domain inside a strip of width $\mathcal{O}(\epsilon)$ around the segment $[X_0(z_0 = \pi), X_0(z_0 = 0)]$ on the x-axis in the narrow gap between the shaft and the casing in the mixing zone.

Introduction of the time modulation breaks the coincidental stable and unstable manifolds (stagnation streamlines) Λ^{z_0} and Γ^{z_0} and causes them to intersect transversely. To be precise, there exist four infinitely long distinguished streaklines (manifolds) which intersect in two intertwined homoclinic tangles, see Figure 7.3.

We label these $\Gamma^S(X_\epsilon(z))$ and $\Lambda^S(X_\epsilon(z))$ because they are the stable manifolds of $X_\epsilon(z)$ and remnants of the branches of the stable manifolds of $X_0(z_0)$ which coincide with Γ^{z_0} and Λ^{z_0}, respectively. The other two are the unstable manifolds of $X_\epsilon(z)$. Although one only observes the unstable manifolds in experiments, one can obtain the stable manifolds from symmetry considerations for special z_0 ($z_0 = 0, \pi$ mod 2π for (MP1) and (MP2)) and in general by performing a second experiment in which the directions of the rotation of the two cylinders is reversed.

Going outward on $\Gamma^U(X_\epsilon(z))$ from $X_\epsilon(z)$ there is a point h_0 at which $\Gamma^U(X_\epsilon(z))$ first intersects $\Gamma^S(X_\epsilon(z))$, (see Figure 7.3). Similarly, leaving from $X_\epsilon(z)$ along the manifold $\Lambda^U(X_\epsilon(z))$, we see that $\Lambda^U(X_\epsilon(z))$ first intersects $\Lambda^S(X_\epsilon(z))$ at a point on the x–axis which we label k_0. We label the segment of $\Gamma^U(X_\epsilon(z))$ between $X_\epsilon(z)$ and h_0 by $U[X_\epsilon(z), h_0]$ and that of $\Gamma^S(X_\epsilon(z))$ between $X_\epsilon(z)$ and h_0 by $S[X_\epsilon(z), h_0]$. Similarly, $U[X_\epsilon(z), k_0]$ denotes the segment of $\Gamma^U(X_\epsilon(z))$ between $X_\epsilon(z)$ and k_0, and $S[X_\epsilon(z), h_0]$ denotes that of $\Gamma^S(X_\epsilon(z))$ between $X_\epsilon(z)$ and k_0. h_0 and k_0 are primary intersection points (pip's).

With the choice of $z_0 = 0 \bmod 2\pi$, the symmetry of the Poincaré map:

$$n \to -n, \qquad x \to x, \qquad y \to -y$$

is such that the unions

$$B_{1,2} \equiv U[X_\epsilon(z), h_0] \bigcup S[X_\epsilon(z), h_0],$$

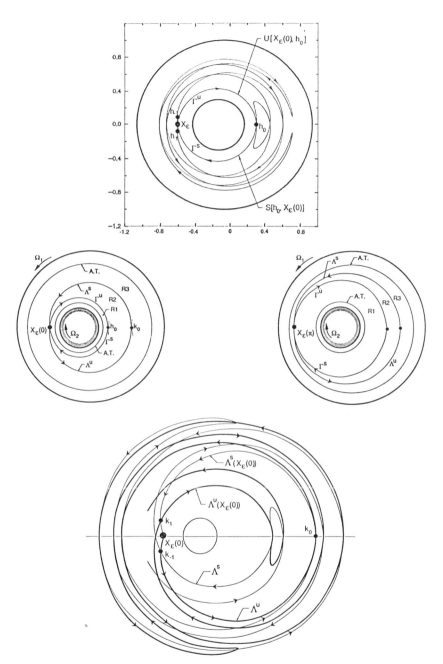

Figure 7.3. $\bar{e} = 0.1$, $\bar{r} = 0.3$ $\epsilon = \frac{2\pi}{40}$, and (MP1). a) Segments of the stable and unstable manifolds, the pip's h_i for $i = -1, 0, 1$. h_1 and h_{-1} are exponentially close (in ϵ) to X_ϵ. b) The regions R1 – R3, for $z = 0$ in the left picture and for $z = \pi$ in the right frame. AT denotes the extremal Arnold torus. The annular regions between the cylinders and these extremal tori (*i.e.*, complementary to R1 – R3) are regular zones filled with tori exponentially close to each other. c) A schematic of the homoclinic tangle formed by Λ^U and Λ^S.

$$B_{2,3} \quad \equiv \quad U[\mathbf{X}_\epsilon(z), k_0] \bigcup S[\mathbf{X}_\epsilon(z), k_0], \tag{7.4}$$

naturally divide the mixing zone into three regions in the time-periodic flow. (See Figure 7.3b.) Region 1 (R1) is the annular domain bounded on the inside by the outermost Arnold torus of the family of tori which make up the regular zone adjacent to the shaft and on the outside by $B_{1,2}$. The kidney-shaped domain in the middle of the mixing zone bounded on the outside by $B_{1,2}$ and $B_{2,3}$ is region 2 (R2). To completely define it, however, we note from Kaper and Wiggins [1993] that there are two possibilities. Either ϵ is large enough such that the entire domain between $B_{1,2}$ and $B_{2,3}$ is part of the actual mixing zone, e.g., $\epsilon \geq 0.14$ in the case when $\bar{e} = 0.1$, $\bar{r} = 0.3$, and one uses (MP2), or ϵ is small and there exists a regular region occupying part of the minimal backflow region. In the former case, the entire kidney-shaped domain is R2, and in the latter case, R2 is an annular domain and the inner boundary is the outermost Arnold torus in the family of tori which make up the regular zone.

Finally, region 3 (R3) is also an annular domain bounded on the inside by $B_{2,3}$ and on the outside by the smallest Arnold torus in the family of tori in the regular zone adjacent to the casing. Thus, the boundaries $B_{1,2}$ and $B_{2,3}$ act as the dividing curve between the three regions. Of course, one can identify three regions and their natural boundaries for every value of z_0 in $[0, 2\pi)$, but for simplicity, we focus only on two choices of z_0: $z_0 = 0, \pi$.

In contrast to the situation in the steady state, transport between the three regions is possible in the modulated flow via the turnstile mechanism. The lobes which are defined by the segments of the manifolds between h_{-1} and h_0 and those between h_{-2} and h_{-1}, as well as the corresponding ones between k_{-1} and k_0 and between k_{-2} and k_{-1} of the Λ-tangle, are the turnstile lobes for this problem. In Figure 7.5 we illustrate the turnstile construction.

Transport in Half Period Intervals

The geometrical mechanisms convecting particles from one region to another are the turnstile lobes. As we stated in the previous subsection both the Γ- and the Λ-tangles have a pair of turnstile lobes. In this and the next section, we prove that the pair of lobes $L_{1,2}(1)$, defined by h_{-2} and h_{-1}, and $L_{2,1}(1)$, defined by h_{-1} and h_0, govern the transport out of and into, respectively, R1, and the pair of lobes $L_{2,3}(1)$ and $L_{3,2}(1)$, defined by the pairs k_{-2} with k_{-1} and k_{-1} with k_0, govern the transport into and out of, respectively, R3.

During the first half of the modulation period, the areas of R1 and R2 increase, while that of R3 decreases. For example, as we see from Figure 7.2, the area enclosed by the instantaneous Γ^z increases as Γ^z sweeps outward away from the shaft for z increasing from 0 to π. Only those tracer particles which are in lobe $L_{2,1}(1)$ (which is a subset of R2 and R3) at time $z = 0 \bmod 2\pi$ enter into R1 in one half of a period. Similarly, only those particles which are in lobe $L_{3,2}(1)$ (which is a subset of R3) at time $z = 0 \bmod 2\pi$ will be in R2 and R1 at time $z = \pi \bmod 2\pi$.

An analogous result is true for the remainder of the modulation period. During the second half of the period, the above is reversed, because the areas of R1 and

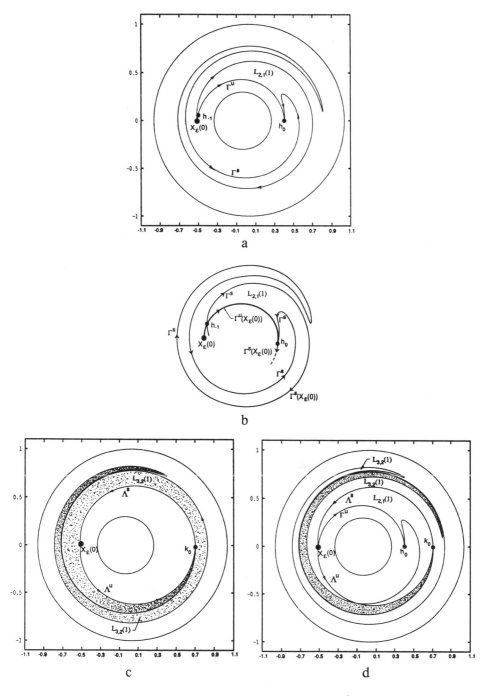

Figure 7.4. The turnstile lobes when $\bar{e} = 0.1$, $\bar{r} = 0.3$, $\epsilon = \frac{2\pi}{40}$, and (MP1) is used. a) The turnstile lobe $L_{2,1}(1)$ of the Γ-tangle. The "tip" of the lobe is in the upper right quadrant of the domain, and the "base" point is the midpoint of the segment of $\Gamma^U(\mathbf{X}_\epsilon)$ between h_{-1} and h_0. b) A schematic of frame a). c) The turnstile lobe $L_{3,2}(1)$ of the Λ-tangle. d) The intersection of $L_{3,2}(1)$ and $L_{2,1}(1)$ is shaded.

R2 decrease, while that of R3 increases. The area enclosed by the instantaneous Γ^z decreases.

The mechanisms by which fluid exits these regions are turnstile lobes. In particular, only those tracer particles which are in lobe $L_{1,2}(1)$ (which is a subset of R1) at time $z = \pi$ mod 2π enter into R2 and R3 during the second half of the modulation period and lie entirely in those regions at time $z = 0$ mod 2π. Similarly, only those particles which are in lobe $L_{2,3}(1)$ (which is a subset of R2 and R1) at time $z = \pi$ mod 2π will lie completely in R3 at time $z = 0$ mod 2π.

To illustrate the above statements, we examine the case in which $\bar{e} = 0.1$, $\bar{r} = 0.3$, $\epsilon = \frac{2\pi}{40}$, and (MP1) is used. (See Figure 7.5.)

We cover the lobe $L_{2,1}(1)$ in R2 (and partially in R3) with a uniform grid of points (spacing=0.006) at time $z = 0$, as shown in Figure 7.5a. We then show in Figure 7.5b that the tracer lies inside R1, exactly in the spiral-shaped turnstile lobe $T_0^{\frac{1}{2}} L_{2,1}(1)$, at time $z = \pi$. In Figures 2.18C and D of Kaper [1991], the shaded region is blown up so that the boundaries of the thin, lamellar striations are clearly visible. Particles only lie in the lobes.

Furthermore, two cases are possible, one in which $L_{2,1}(1) \cap L_{3,2}(1) \neq \emptyset$, as shown in Figure 7.4d, and the other in which $L_{2,1}(1) \cap L_{3,2}(1) = \emptyset$. However, for $\epsilon < 0.3$ in all of the cases we analyzed, the former holds true. This explains why we said above that $L_{2,1}(1)$ is a subset of R2 and R3. The results we present in the remainder of this chapter apply in both cases; one need only be slightly careful about justifying the formulae we use in the former case, as we show in the next section.

Transport in Intervals of Unit Periods

In this section, we look at the inter-regional transport process from the usual, per-unit-period point of view and obtain results giving the probability that an orbit initially in one region can be found in another region after any period of the modulation. Rather than treat this problem in its full generality immediately, however, we first consider the particular problem of what fraction of tracer initially in R3 gets transported to R1 in each period of the modulation.

One may then compute the other eight quantities $T_{i,j}(n)$ for $i,j = 1,2,3$, which, assuming that region i is initially filled with tracer fluid, represent the amount of tracer initially in region i that is in region j exactly at the end of the n-th period. These can be used to give the probabilities, using the same procedure as we do here for $T_{3,1}(n)$ and the conservation equations. These equations, five in all for the nine independent quantities $T_{i,j}(n)$ for $i,j = 1,2,3$, express the conservation of tracer and the conservation of the areas $\mu(R_i)$ for $i = 1,2,3$:

$$\sum_{j=1}^{3}(T_{i,j}(n) - T_{i,j}(n-1)) = 0 \qquad i = 1,2,3$$

$$\sum_{i=1}^{3}(T_{i,j}(n) - T_{i,j}(n-1)) = 0 \qquad j = 1,2,3. \qquad (7.5)$$

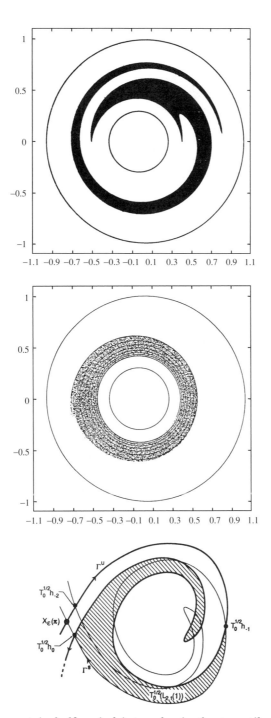

Figure 7.5. Transport in half-period intervals via the turnstile mechanism in the case $\bar{e} = 0.1$, $\bar{r} = 0.3$ $\epsilon = \frac{2\pi}{40}$, and (MP1). a) All tracer particles at $z = 0$ lie in lobe $L_{2,1}(1)$, which is shaded. b) At $z = \pi$, all tracer particles lie in R2, in the half-period image of the lobe $L_{2,1}(1)$. Enlargement of frame b) shows that all of the particles lie inside the half-period image of the lobe, which has a solid boundary. c) Schematic of frame b).

The solution to this problem represents the probability that an orbit, initially rotating in the same sense as the casing, changes the direction in which it is flowing in as a function of the modulation period. To define the problem precisely, we assume that the tracer is uniformly concentrated in R3 initially, *i.e.*, at the slow time $z = 0$. The question we answer, then, is: How much tracer is in R1 at time $z = 2n\pi$ for $n = 1, 2, ...$? We label this quantity as $T_{3,1}(n)$.

Before proceeding, we must redefine the turnstile lobes to eliminate the intra-turnstile overlap areas $L_{1,2}(1) \cap L_{2,1}(1)$ and $L_{2,3}(1) \cap L_{3,2}(1)$. The two lobes forming a turnstile always intersect in adiabatic dynamical systems due to the lobe area theorem given in Kaper and Wiggins [1992]. In particular, we set

$$
\begin{aligned}
\tilde{L}_{1,2}(1) &\equiv L_{1,2}(1) \cap \text{R1}, \\
\tilde{L}_{2,1}(1) &\equiv L_{2,1}(1) - L_{2,1}(1) \cap L_{1,2}(1), \\
\tilde{L}_{2,3}(1) &\equiv L_{2,3}(1) \cap (\text{R2} \cup \text{R1}), \\
\tilde{L}_{3,2}(1) &\equiv L_{3,2}(1) - L_{3,2}(1) \cap L_{2,3}(1).
\end{aligned}
\tag{7.6}
$$

The excluded parts of the original turnstile lobes, although they get mapped across the inter-regional boundaries during the first half of the modulation period, get mapped back across before the end of the period to the region they were in at the beginning of the period. For example, during every period, the fluid in $L_{2,1}(1) \cap L_{1,2}(1)$ gets mapped from R2 into R1 and back again. Therefore, the orbits in these overlapping regions change their orbit type (*i.e.*, rotate in the same sense as the shaft or the casing, or are in the instantaneous backflow region) an even number of times in each period, and the parts in the redefined lobes do so an odd number of times.

Now, the fact that Γ^S cannot self-intersect implies that $L_{1,2}(1)$ must lie in $L_{2,1}(1)$ and R1, because as soon as it crosses Γ^U between $\mathbf{X}_\epsilon(z_0)$ and h_1, it does so at the pip which defines the boundary of $L_{1,2}(1)$. Thus, the overlap region is large. In fact, the turnstile lobes overlap completely in the limit of $\epsilon \to 0$, as we will see in the next section.

In addition, exclusion of the intra-turnstile overlap, which may be characterized by saying that we exclude, for example, from $L_{2,1}(1)$ the (large) piece of the lobe $L_{1,2}(1)$ "nested" inside $L_{2,1}(1)$, isolates the long, thin, folded structure of $\tilde{L}_{2,1}(1)$ that gets transported into R1 at the end of each cycle. We now show that it is responsible for continuing the stretching and folding of the lamellar tracer structure discussed in the previous section as z increases from $z = \pi$ to $z = 2\pi$ and in each subsequent period.

$\tilde{L}_{2,1}(1)$ directly transports an amount of fluid equal to $\mu(\tilde{L}_{2,1}(1) \cap \tilde{L}_{3,2}(1))$ from R3 into R1 during each period, where $\mu(L)$ denotes the area of the planar set L. In fact, only the tracer contained in the intersections of $\tilde{L}_{2,1}(1)$ with $T_0^{n-k}\tilde{L}_{3,2}(1)$, for $k = 1, 2, ..., n-1, n$, can enter R1 at the n-th iteration.

However, the intersection $\tilde{L}_{2,1}(1) \cap \tilde{L}_{3,2}(1)$ is only uniformly filled with tracer fluid initially. At later times, the concentration of tracer is not uniform. Furthermore, although the fluid in this and the other intersections is exactly what we need, not all of it is tracer fluid. Thus, the problem requires us not only to identify the flux mechanism, as we have done so far, but also to find a way to determine the content of the lobes.

Having redefined the turnstile lobes, the transport theory presented in Rom-Kedar and Wiggins [1990] directly determines the lobe content. The two main quantities needed are the amount of tracer which is in the lobes $\tilde{L}_{2,1}(n)$ and $\tilde{L}_{1,2}(n)$, which we denote by $\tilde{L}^3_{2,1}(n)$ and $\tilde{L}^3_{1,2}(n)$ following the notation of Rom-Kedar and Wiggins [1990], where the superscript 3 indicates that we are following the tracer which was initially uniformly distributed in R3. For $n > 1$, we find

$$\mu\left(\tilde{L}^3_{2,1}(n)\right) = \mu\left(\tilde{L}_{2,1}(1) \cap \tilde{L}_{3,2}(1)\right)$$

$$+ \sum_{k=1}^{n-1} \left\{ \mu\left(T_0^{k-n}\tilde{L}_{2,1}(1) \cap L_{3,2}(1)\right) - \mu\left(T_0^{k-n}\tilde{L}_{2,1}(1) \cap \tilde{L}_{2,3}(1)\right) \right\},$$

$$\mu\left(\tilde{L}^3_{1,2}(n)\right) = \mu\left(\tilde{L}_{1,2}(1) \cap \tilde{L}_{3,2}(1)\right)$$
(7.7)

$$+ \sum_{k=1}^{n-1} \left\{ \mu\left(T_0^{k-n}\tilde{L}_{1,2}(1) \cap L_{3,2}(1)\right) - \mu\left(T_0^{k-n}\tilde{L}_{1,2}(1) \cap \tilde{L}_{2,3}(1)\right) \right\}.$$

At first glance it may seem that, in addition to redefining the turnstile lobes as we did above, we also must modify the transport theory, because of the fact that part of $\tilde{L}_{2,1}(1)$ lies in R3 and part of $\tilde{L}_{2,3}(1)$ lies in R1. However, the above formulae are exactly the necessary ones because the intersections with $\tilde{L}_{3,2}(1)$ and $\tilde{L}_{2,3}(1)$ are accounted for from the first period onward.

Next, the change in the amount of tracer in R1 at the n-th cycle is

$$T_{3,1}(n) - T_{3,1}(n-1) = \mu\left(\tilde{L}^3_{2,1}(n)\right) - \mu\left(\tilde{L}^3_{1,2}(n)\right).$$
(7.8)

Using (7.7), we evaluate the right hand side. Finally, we write a (telescopically-collapsing) sum using the above difference formula to obtain $T_{3,1}(n)$ strictly as a function of $T_{3,1}(0)$ and the lobe content expressions (7.7):

$$T_{3,1}(n) = T_{3,1}(0)$$

$$+ \sum_{m=1}^{n-1}(n-m) \left\{ \mu\left(\tilde{L}_{2,1}(1) \cap T_0^m\tilde{L}_{3,2}(1)\right) - \mu\left(\tilde{L}_{2,1}(1) \cap T_0^m\tilde{L}_{2,3}(1)\right) \right\}.$$
(7.9)

Since, by assumption, all of the tracer is in R3 initially, $T_{3,1}(0)$ is identically zero, and (7.9) reduces to

$$T_{3,1}(n) = \sum_{m=1}^{n-1}(n-m) \left\{ \mu\left(\tilde{L}_{2,1}(1) \cap T_0^m\tilde{L}_{3,2}(1)\right) - \mu\left(\tilde{L}_{2,1}(1) \cap T_0^m\tilde{L}_{2,3}(1)\right) \right\}.$$
(7.10)

7.4. Analytical Estimates from Adiabatic Dynamical Systems Theory

Adiabatic Melnikov Theory

In this section, we demonstrate the use of the adiabatic Melnikov function, $M_A(z)$, described in the previous chapter. Recall that it is the coefficient of the leading order term in an asymptotic series for the distance between the stable and unstable manifolds forming a "same pair" tangle as measured along the normal to the instantaneous stagnation streamlines, and is given by

$$M_A(z) \equiv \int_{-\infty}^{\infty} t \cdot \left(\frac{\partial \psi}{\partial x} \frac{\partial^2 \psi}{\partial y \partial z} - \frac{\partial \psi}{\partial y} \frac{\partial^2 \psi}{\partial x \partial z} \right) (x_0^z(t), y_0^z(t); z) dt, \qquad (7.11)$$

where $(x_0^z(t), y_0^z(t))$ is an orbit parametrizing the instantaneous stagnation streamline for which the function is being evaluated, either Γ^z or Λ^z. If this integral is evaluated along a homoclinic orbit on Γ^z, we denote the adiabatic Melnikov function by $M_A^\Gamma(z)$, and it measures the distance between Γ^S and Γ^U. Similarly, we write $M_A^\Lambda(z)$ when the function is evaluated along an orbit on Λ^z.

A simple zero of $M_A^\Gamma(z)$ implies that, for ϵ sufficiently small, Γ^S and Γ^U intersect each other transversely. Likewise for $M_A^\Lambda(z)$ and the manifolds Λ^S and Λ^U. Furthermore, for periodic and quasiperiodic systems, one intersection of the manifolds implies that there are infinitely many others, because, as we stated in the previous section, invariance of the manifolds implies that a point on both manifolds must always be on both manifolds.

The theory presented in Kaper and Wiggins [1991a] shows that $M_A^\Gamma(z) = \frac{dA^\Gamma}{dz}(z, z_0)$, where A^Γ is the difference between the areas enclosed by the instantaneous separatrices Γ^z and Γ^{z_0} and z_0 is the zero of $M_A^\Gamma(z)$ corresponding to the nearest intersection and extremal instantaneous separatrix. A similar result holds for $M_A^\Lambda(z)$. Thus, since $\frac{dA}{dz}$ changes sign at the extremal values of z, we know that both $M_A^\Gamma(z)$ and $M_A^\Lambda(z)$ have periodically spaced simple zeroes for our modulation protocols at $z = 0, \pi \mod 2\pi$, and hence the intersecting stagnation streamlines form two homoclinic tangles as shown in the figures of the previous section. Thus, we have rigorously established the existence of the intersections of the manifolds we obtained numerically in the previous section.

Furthermore, the theory of the adiabatic Melnikov function enables us to show that all of the pip's h_i and k_i, for the integer i except $i = 0$, lie in a small neighborhood, $\mathcal{N}_\epsilon$, whose size depends on, of ϵ, of $\mathbf{X}_\epsilon(z)$ on the Poincaré section. Since the z–distance between adjacent zeros of both $M_A^\Gamma(z)$ and $M_A^\Lambda(z)$ is equal to π, we know that in the fast time t, two adjacent pip's are separated from each other by a time-of-flight of $\Delta t = \frac{\pi}{\epsilon}$ along Γ^z and Λ^z, respectively. Thus, on the Poincaré section with $z = 0$, all of the pip's h_i and k_i, except h_0 and k_0 which lie near the respective reference points on Γ^z and Λ^z, lie exponentially close in time to $\mathbf{X}_\epsilon(z)$ and, hence, in $\mathcal{N}_\epsilon$. This, in turn, implies that the tangles are as shown in the figures from the previous section and that they are very difficult to obtain accurately by numerical methods. Even for values of ϵ as large as $\epsilon = \frac{2\pi}{20}$, the time-of-flight is large enough so that all of the pip's but h_0 and k_0 lie in $\mathcal{N}_\epsilon$. Also, the manifolds are making room for the growing regular zone, which is made up of an increasingly larger number of persistent Arnold tori as $\epsilon \to 0$, in the instantaneous backflow region.

Finally, (7.11) may be rewritten in a computationally more convenient manner, as is shown in Kaper and Wiggins [1992], as

$$M_A(z) = \int_{-\infty}^{\infty} \left(\frac{\partial \psi}{\partial z}(\Gamma(z), z) - \frac{\partial \psi}{\partial z}(\mathbf{X}_0^z, z) \right) dt. \qquad (7.12)$$

Using this form, one can see that the adiabatic Melnikov functions for both separatrices have simple zeros at $z = 0, \pi$ mod 2π, because the derivative of ψ only contains terms proportional to $\sin z$, and hence vanishes there.

Lobe Area

The second result we can use is a formula for lobe area. The formula proven in Kaper and Wiggins [1992] states that the area of a lobe is given to leading order by the difference between the areas enclosed by the maximum and minimum instantaneous stagnation streamlines that occur during the modulation period. Thus, to leading order, the area is an $\mathcal{O}(1)$ quantity as $\epsilon \to 0$, which is strikingly different from the regular perturbation case in which the leading order term is $\mathcal{O}(\delta)$ asymptotically. To be more specific, the leading order term in the asymptotic expansion for the area of $L_{2,1}(1)$ and all of the other lobes in the Γ-tangle (*i.e.* the tangle formed by the transverse intersections of Γ^S and Γ^U) is the difference between the maximum and minimum areas of region A attained during the given protocol. Similarly, the leading order term in the asymptotic expansion for the area of $L_{3,2}(1)$ and all of the other lobes in the Λ-tangle is the difference between the maximum and minimum areas of region C attained during the given protocol.

The adiabatic Melnikov function can be used to obtain an approximation of the lobe area. In Kaper and Wiggins [1992] it has been shown that the integral of $M_A(z)$ between two of its adjacent simple zeros gives the same result as the leading order term for the lobe area. Furthermore, the remaining terms in the asymptotic expansion are $\mathcal{O}(\epsilon)$.

CHAPTER 8

Orbits Homoclinic to Resonances: Global Analysis and Geometric Singular Perturbation Methods

8.1. The Basic Perturbed, Two-Degree-of-Freedom Integrable Hamiltonian System

The general systems under consideration have the following form:

Non-Hamiltonian Perturbations

$$
\begin{aligned}
\dot{x} &= JD_x H(x, I; \mu) + \epsilon g^x(x, I, \theta; \mu, \epsilon), \\
\dot{I} &= \epsilon g^I(x, I, \theta; \mu, \epsilon), \qquad\qquad (x, I, \theta) \in \mathbb{R}^2 \times \mathbb{R} \times S^1, \\
\dot{\theta} &= D_I H(x, I; \mu) + \epsilon g^\theta(x, I, \theta; \mu, \epsilon),
\end{aligned}
\tag{8.1}
$$

Hamiltonian Perturbations

$$
\begin{aligned}
\dot{x} &= JD_x H(x, I; \mu) + \epsilon JD_x H_1(x, I, \theta, \mu, \epsilon), \\
\dot{I} &= -\epsilon D_\theta H_1(x, I, \theta, \mu, \epsilon), \qquad (x, I, \theta) \in \mathbb{R}^2 \times \mathbb{R} \times S^1, \\
\dot{\theta} &= D_I H(x, I; \mu) + \epsilon D_I H_1(x, I, \theta, \mu, \epsilon),
\end{aligned}
\tag{8.2}
$$

where J denotes the usual symplectic matrix, i.e.,

$$
\begin{pmatrix} 0 & 1 \\ -1 & 0 \end{pmatrix},
$$

all functions are sufficiently differentiable (C^r, $r \geq 3$ is sufficient) on the domains of interest, $0 < \epsilon << 1$ is the perturbation parameter, and $\mu \in V \subset \mathbb{R}^p$ is a vector of parameters. (Note: D_x, etc. will denote partial derivatives and $\frac{d}{dx}$, etc. will denote total derivatives.)

The phenomena that we encounter will be very different for the cases of Hamiltonian and non-Hamiltonian perturbations; which is why we must treat them differently. The main point behind these global perturbation methods is to use the geometrical structure of the integrable Hamiltonian unperturbed problem in order to develop appropriate "coordinates" for studying the perturbed problem. Hence, we begin by stating our assumptions on the unperturbed problem.

The techniques described in the following were developed in collaboration with Gregor Kovačič for the case of non-Hamiltonian perturbations and orbits homoclinic to equilibrium points of saddle-focus type. The techniques for non-Hamiltonian perturbations and orbits homoclinic to equilibrium points of saddle-saddle type were worked out in collaboration with Dave McLaughlin, and were greatly aided by the numerical experiments of Ed Overman and X. Xiong. The Hamiltonian methods were all done in collaboration with György Haller. In fact, this chapter is a summary of work found in Kovačič and Wiggins [1992], Haller and Wiggins [1993a], and McLaughlin *et al.* [1993].

8.2.   The Analytic and Geometric Structure of the Unperturbed Equations

The unperturbed equations are given by

$$
\begin{aligned}
\dot{x} &= JD_xH(x,I;\mu), \\
\dot{I} &= 0, \qquad\qquad (x,I,\theta;\mu) \in \mathbb{R}^2 \times \mathbb{R} \times S^1, \\
\dot{\theta} &= D_IH(x,I;\mu).
\end{aligned}
\tag{8.3}
$$

Note the simple structure of (8.3); effectively, they are two uncoupled one-degree-of-freedom Hamiltonian systems. Since $\dot{I} = 0$, the I variable enters the x component of (8.3) only as a parameter. Hence, the x component of (8.3) can be solved exactly since it is just a one parameter family of one-degree-of-freedom (hence integrable) Hamiltonian systems. This solution can then be substituted into the θ component of (8.3), which can then be integrated to yield the full solution. We make the following assumption on the x component of (8.3):

Assumption 1. *For all $I \in [I_1, I_2]$, $\mu \in V$, the equation*

$$
\dot{x} = JD_xH(x,I;\mu)
\tag{8.4}
$$

has a hyperbolic fixed point, $\tilde{x}_0(I;\mu)$, connected to itself by a homoclinic orbit, $x^h(t,I;\mu)$, i.e., $\lim_{t\to\pm\infty} x^h(t,I;\mu) = \tilde{x}_0(I;\mu)$.

We now want to use the simple structure of the "decoupled" unperturbed system to build up a picture of the geometry in the full four-dimensional phase space. This will provide the framework for studying the perturbed problem, which will be fully four-dimensional.

Assumption 1 implies that in the full four-dimensional phase space, the set

$$
\mathcal{M} = \{(x,I,\theta) \mid x = \tilde{x}_0(I;\mu),\ I_1 \le I \le I_2,\ 0 \le \theta < 2\pi,\ \mu \in V\}
\tag{8.5}
$$

is a two-dimensional, invariant manifold. Moreover, the hyperbolic saddle stability of the fixed point from Assumption 1 gives rise to *normal hyperbolicity* of $\mathcal{M}$. Normal hyperbolicity is a technical term that means that, under the linearized dynamics, rates of expansion and contraction transverse to $\mathcal{M}$ dominate those tangent to $\mathcal{M}$; it is important since, normally, hyperbolic invariant manifolds persist under perturbation. Further information related to the normal hyperbolicity of $\mathcal{M}$ can be found in Wiggins [1988a], [1993].

The two-dimensional normally hyperbolic invariant manifold $\mathcal{M}$ has three- dimensional stable and unstable manifolds, which we denote as $W^s(M)$ and $W^u(M)$, respectively. This can be inferred from the structure of the x component of (8.3) given in Assumption 1, as well as from the proof of normal hyperbolicity of $\mathcal{M}$ given in Wiggins [1988a], [1993]. Moreover, the existence of the homoclinic orbit of the x component of (8.3) implies that $W^s(M)$ and $W^u(M)$ intersect (nontransversely) along a three-dimensional *homoclinic manifold* which we denote by Γ. More precisely, we have

$$\Gamma \equiv W^s(M) \cap W^u(M)$$

$$= \{(x, I, \theta) \mid H(x, I; \mu) - H(\tilde{x}_0(I; \mu), I; \mu) = 0, \ I_1 \le I \le I_2, \ 0 \le \theta < 2\pi, \ \mu \in V\}.$$

A trajectory in $\Gamma \equiv W^s(M) \cap W^u(M)$ can be expressed as

$$\left(x^h(t, I; \mu), \ I, \ \theta(t, I, \theta_0; \mu) = \int_0^t D_I H(x^h(s, I; \mu), I; \mu)ds + \theta_0\right), \qquad (8.6)$$

and it is clear that this trajectory approaches an orbit in $\mathcal{M}$ as $t \to \pm\infty$ since $x^h(t, I; \mu) \to \tilde{x}_0(I; \mu)$ as $t \to \pm\infty$.

8.2.1. The Dynamics of the Unperturbed System Restricted to $\mathcal{M}$. The unperturbed system restricted to $\mathcal{M}$ is given by

$$\begin{aligned} \dot{I} &= 0, \\ \dot{\theta} &= D_I H(\tilde{x}_0(I; \mu), I; \mu), \qquad\qquad I_1 \le I \le I_2. \qquad (8.7) \end{aligned}$$

Thus if $D_I H(\tilde{x}_0(I; \mu), I; \mu) \ne 0$ then $I = $ constant labels a periodic orbit and if $D_I H(\tilde{x}_0(I; \mu), I; \mu) = 0$ then $I = $ constant labels a circle of fixed points. We refer to a value of I for which $D_I H(\tilde{x}_0(I; \mu), I; \mu) = 0$ as a *resonant I value* and these fixed points as *resonant fixed points* and we make the following assumption on the unperturbed system restricted to $\mathcal{M}$:

Assumption 2–Resonance. *There exists a value of $I \in [I_1, I_2]$, denoted I^r, at which $D_I H(\tilde{x}_0(I^r; \mu), I^r; \mu) = 0$.*

8.2.2. The Dynamics in Γ and its Relation to the Dynamics in $\mathcal{M}$. Recall the expression for an orbit in Γ given in (8.6). As $x^h(t, I; \mu) \to \tilde{x}_0(I; \mu)$ and I remains constant, we want to call attention to the expression that we will define as

$$\Delta\theta(\mu) = \theta(+\infty, I, \theta_0; \mu) - \theta(-\infty, I, \theta_0; \mu),$$

$$= \int_{-\infty}^{+\infty} D_I H\left(x^h(t, I; \mu), I; \mu\right) dt. \qquad (8.8)$$

Now, for I such that $D_I H(\tilde{x}_0(I; \mu), I; \mu) \neq 0$, it is easy to see that $\Delta\theta(\mu)$ is not defined. This just reflects the fact that, asymptotically, the orbit approaches a periodic orbit whose phase constantly changes forever. However, at resonant I values, $\Delta\theta$ is defined since the integral converges (convergence of the integral follows from the fact that $x^h(t, I; \mu) \to \tilde{x}_0(I; \mu)$ exponentially fast as $t \to \pm\infty$, hence at resonance $D_I H(x^h(t, I; \mu), I; \mu)$ goes to zero exponentially fast as t goes to $\pm\infty$). Since resonant I values denote circles of fixed points on $\mathcal{M}$ the orbit $(x^h(t, I^r; \mu), I^r, \theta(t, I^r, \theta_0; \mu))$ is typically a heteroclinic connection between different points on the resonant circle of fixed points (the connection will be homoclinic if $\Delta\theta(\mu) = 2\pi n$, for some integer n). The number $\Delta\theta(\mu)$ gives the shift in phase between the two endpoints of the heteroclinic trajectory along the circle of fixed points.

In Figure 8.1 we illustrate the relevant aspects of the geometry and dynamics of the unperturbed system. The three parts of this figure convey the essence of the rest of this chapter. In Figure 8.1a we see the three-dimensional stable and unstable manifolds of the hyperbolic invariant set. *As manifolds*, these are affected little by the perturbation (neglecting intersection properties). In Figure 8.1b we see the resonant "elliptic" dynamics on $\mathcal{M}$. This is affected *drastically* by the perturbation. Finally, in Figure 8.1c, we see connections between orbits in the stable and unstable manifolds of $\mathcal{M}$ to orbits on $\mathcal{M}$. This is precisely the issue we will study in the perturbed problem.

8.3.   The Analytic and Geometric Structure of the Perturbed Equations

The geometrical structure of the unperturbed system will provide us with the framework for understanding certain types of global behavior that can occur in the perturbed system. In particular, $\mathcal{M}$, along with its stable and unstable manifolds will persist in the perturbed system (under arbitrary, sufficiently differentiable perturbations); however, the dynamics on these manifolds will be quite different in the perturbed system. This should be evident since $\mathcal{M}$ contains a circle of fixed points and the stable and unstable manifolds of $\mathcal{M}$ have a coinciding branch. Nevertheless, these geometric structures will guide our analysis, in that, we will study the nature of the dynamics on the perturbed $\mathcal{M}$ near the resonance (this is where the Hamiltonian vs. non-Hamiltonian nature of the perturbation is crucial), then we will study the influence of the perturbation on the stable and unstable manifolds of $\mathcal{M}$, and, finally, we will relate these two aspects of the dynamics for the perturbed system.

8.3.1.   The Persistence of $\mathcal{M}$ and its Stable and Unstable Manifolds.
The question of the persistence under perturbations of invariant manifolds with boundary gives rise to certain technical questions concerning the nature of the trajectories at the boundary. In order to address these questions precisely, we begin by defining the following set:

$$U^\delta = \{(x, I, \theta) \mid |x - \tilde{x}_0(I; \mu)| \leq \delta, \tilde{I}_1 \leq I \leq \tilde{I}_2\}, \tag{8.9}$$

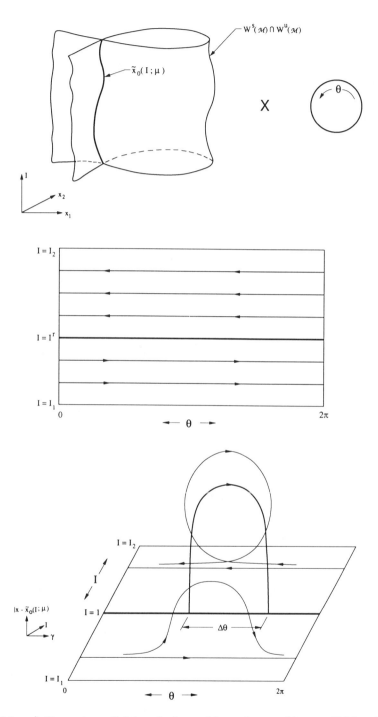

Figure 8.1. a) Geometry of $\mathcal{M}$ and the stable and unstable manifolds of $\mathcal{M}$. b) The dynamics on $\mathcal{M}$. c) The geometry of trajectories homoclinic to the periodic orbits on $\mathcal{M}$ and orbits heteroclinic to fixed points on the resonant circle of fixed points.

where

$$I_1 \leq \tilde{I}_1 < \tilde{I}_2 \leq I_2.$$

If $I_1 = \tilde{I}_1$ and $I_2 = \tilde{I}_2$ then clearly U^δ is a neighborhood of $\mathcal{M}$. However, for technical reasons (to be discussed shortly), we may need to slightly restrict the range of I values in discussing the perturbed manifolds, and it is for this reason that the I interval in the definition of U^δ has been restricted. (Note: it will turn out that $\tilde{I}_1$ can be chosen arbitrarily close to I_1 and $\tilde{I}_2$ can be chosen arbitrarily close to I_2.) The set U^δ will be useful in characterizing the nature of trajectories near the invariant manifolds. For the unperturbed system, we define the local stable and unstable manifolds of $\mathcal{M}$ as follows:

$$W_{loc}^s(\mathcal{M}) \equiv W^s(\mathcal{M}) \cap U^\delta, \tag{8.10}$$
$$W_{loc}^u(\mathcal{M}) \equiv W^u(\mathcal{M}) \cap U^\delta. \tag{8.11}$$

We now state the persistence theorem which applies to both (8.1) and (8.2).

Theorem 8.3.1. *There exists $\epsilon_0 > 0$ sufficiently small such that for $0 < \epsilon < \epsilon_0$, $\mathcal{M}$ persists as a C^r, locally invariant two-dimensional normally hyperbolic manifold with boundary, which we denote by $\mathcal{M}_\epsilon$, having the following properties:*

1. *$\mathcal{M}_\epsilon$ is C^r in ϵ and μ.*

2. *$\mathcal{M}_\epsilon$ is C^r ϵ-close to $\mathcal{M}$ and can be represented as a graph over $\mathcal{M}$ as follows:*

   $$\mathcal{M}_\epsilon = \{(x, I, \theta) \mid x = \tilde{x}_\epsilon(I, \theta, \mu) = \tilde{x}_0(I; \mu) + \epsilon \tilde{x}_1(I, \theta, \mu) + \mathcal{O}(\epsilon^2), \tilde{I}_1 \leq I \leq \tilde{I}_2\}. \tag{8.12}$$

   Moreover, there exists δ_0 sufficiently small (depending on ϵ) such that for $0 < \delta < \delta_0$ there exists locally invariant manifolds in U^δ, denoted $W_{loc}^s(\mathcal{M}_\epsilon)$, $W_{loc}^u(\mathcal{M}_\epsilon)$, having the following properties.

3. *$W_{loc}^s(\mathcal{M}_\epsilon)$ and $W_{loc}^u(\mathcal{M}_\epsilon)$ are C^r in ϵ and μ.*

4. *$W_{loc}^s(\mathcal{M}_\epsilon) \cap W_{loc}^u(\mathcal{M}_\epsilon) = \mathcal{M}_\epsilon$.*

5. *$W_{loc}^s(\mathcal{M}_\epsilon)$ (respectively $W_{loc}^u(\mathcal{M}_\epsilon)$) is a graph over $W_{loc}^s(\mathcal{M})$ (respectively $W_{loc}^u(\mathcal{M})$) and is C^r ϵ-close to $W_{loc}^s(\mathcal{M})$ (respectively $W_{loc}^u(\mathcal{M})$).*

6. *Let $y_\epsilon^s(t) \equiv (x_\epsilon^s(t), I_\epsilon^s(t), \theta_\epsilon^s(t))$ (respectively $y_\epsilon^u(t) \equiv (x_\epsilon^u(t), I_\epsilon^u(t), \theta_\epsilon^u(t))$) denote a trajectory that is in $W_{loc}^s(\mathcal{M}_\epsilon)$ (respectively $W_{loc}^u(\mathcal{M}_\epsilon)$) at $t = 0$. Then as $t \to +\infty$ $(t \to -\infty)$ either,*

   (a) *$y_\epsilon^s(t) \equiv (x_\epsilon^s(t), I_\epsilon^s(t), \theta_\epsilon^s(t))$ (respectively $y_\epsilon^u(t) \equiv (x_\epsilon^u(t), I_\epsilon^u(t), \theta_\epsilon^u(t))$) crosses ∂U^δ*

   or,

(b) $| y_{\epsilon}^{s}(t) - \mathcal{M}_{\epsilon} | \to 0$ *(respectively* $| y_{\epsilon}^{u}(t) - \mathcal{M}_{\epsilon} | \to 0$*).*

We refer to $W_{loc}^{s}(\mathcal{M}_{\epsilon})$ *and* $W_{loc}^{u}(\mathcal{M}_{\epsilon})$ *as the local stable and unstable manifolds of* $\mathcal{M}_{\epsilon}$*, respectively.*

Proof: The proof follows from the invariant manifold theory developed by Fenichel [1971]. Fenichel's theory is adapted to systems of the form of (8.1) and (8.2) in Wiggins [1988a], [1993] where the details of the proof of this theorem can be found.

□

We make the following remarks concerning the consequences and implications of this theorem:

Remark 1. The term *locally invariant* means that trajectories with initial conditions on $\mathcal{M}_{\epsilon}$ may leave $\mathcal{M}_{\epsilon}$; however, they may do so only by crossing the boundary of $\mathcal{M}_{\epsilon}$. In proving the persistence of $\mathcal{M}$ under perturbation, it is necessary to know the stability properties of trajectories in $\mathcal{M}$ on semi-infinite time intervals. Technically, this control is accomplished by modifying the perturbed vector field (8.1) in an arbitrarily small neighborhood of the boundary of $\mathcal{M}$ by using C^{∞} "bump functions"; this procedure is thoroughly explained in Wiggins [1988a]. The perturbed manifold is then constructed as a graph over the unperturbed manifold by using the graph transform technique. This is the reason why the range of I values for which $\mathcal{M}_{\epsilon}$ exists in the perturbed vector field (8.1) may need to be slightly decreased.

Remark 2. We define the global stable and unstable manifolds of $\mathcal{M}_{\epsilon}$, denoted $W^{s}(\mathcal{M}_{\epsilon})$ and $W^{u}(\mathcal{M}_{\epsilon})$, respectively, as follows: Let $\phi_{t}(\cdot)$ denote the flow generated by (8.1), then we define

$$W^{s}(\mathcal{M}_{\epsilon}) = \bigcup_{t \leq 0} \phi_{t}\left(W_{loc}^{s}(\mathcal{M}_{\epsilon}) \cap U^{\delta}\right),$$

$$W^{u}(\mathcal{M}_{\epsilon}) = \bigcup_{t \geq 0} \phi_{t}\left(W_{loc}^{u}(\mathcal{M}_{\epsilon}) \cap U^{\delta}\right). \tag{8.13}$$

Remark 3. The phrase *stable manifold of an invariant set* typically means the manifold of trajectories that approach the invariant set as $t \to +\infty$. However our definition has a slightly different meaning that is peculiar to our invariant set, i.e., $\mathcal{M}_{\epsilon}$, having a boundary. This is characterized in terms of the alternatives (a) and (b) of part 6 of Theorem 8.3.1; similarly for the *unstable manifold of an invariant set.*

Approximate Calculation of $\mathcal{M}_{\epsilon}$ Near Resonance

The fact that $\mathcal{M}_{\epsilon}$ is C^{r} $(r \geq 2)$ in μ and ϵ allows us to Taylor expand the manifold in powers of μ and ϵ. This will be important because we may need to explicitly compute the $\mathcal{O}(\epsilon)$ term in the expansion of $\tilde{x}_{\epsilon}(I, \theta; \mu)$ given in (8.12). We now explain how this may be done.

The following calculations are independent of whether or not the perturbation is Hamiltonian. Therefore, we give the calculations using the notation of a general perturbation.

Differentation of $\tilde{x}_\epsilon(I, \theta; \mu)$ along the perturbed vector field (8.1) gives a quasi-linear partial differential equation that $\tilde{x}_\epsilon(I, \theta; \mu)$ must satisfy. This equation is given by

$$JD_x H(\tilde{x}_\epsilon, I; \mu) + \epsilon g^x(\tilde{x}_\epsilon, I, \theta; \mu, \epsilon)$$

$$= \epsilon \left(D_I \tilde{x}_\epsilon\right) g^I(\tilde{x}_\epsilon, I, \theta; \mu, \epsilon) + \left(D_\theta \tilde{x}_\epsilon\right) \left(D_I H(\tilde{x}_\epsilon, I; \mu) + \epsilon g^\theta(\tilde{x}_\epsilon, I, \theta; \mu, \epsilon)\right).$$
$$(8.14)$$

We can differentiate (8.14) with respect to ϵ and obtain equations that the derivatives of $\tilde{x}_\epsilon(I, \theta; \mu)$ must satisfy. In this way we find that $\tilde{x}_1(I, \theta; \mu)$ must satisfy the following ordinary differential equation

$$- \left(D_\theta \tilde{x}_1\right) D_I H(\tilde{x}_0(I; \mu), I; \mu) + JD_x^2 H(\tilde{x}_0(I; \mu), I; \mu)\tilde{x}_1$$
$$(8.15)$$
$$= \left(D_I \tilde{x}_0(I; \mu)\right) g^I(\tilde{x}_0(I), I, \theta; \mu, 0) - g^x(\tilde{x}_0(I; \mu), I, \theta; \mu, 0).$$

We want to point out that in one special circumstance (indeed, the situation that will be most important to us), the solution of (8.15) can immediately be written down, namely, at resonance. At resonance, i.e., $I = I^r$, we have $D_I H(\tilde{x}_0(I^r; \mu), I^r; \mu) = 0$ so that (8.15) reduces to an algebraic equation with solution

$$\tilde{x}_1 = \left(JD_x^2 H(\tilde{x}_0(I^r; \mu), I^r)\right)^{-1} \left(D_I \tilde{x}_0(I^r; \mu) g^I(\tilde{x}_0(I^r; \mu), I^r, \theta; \mu, 0)\right.$$
$$(8.16)$$
$$\left. - g^x(\tilde{x}_0(I^r; \mu), I^r, \theta; \mu, 0)\right).$$

It is also easy to find an expression for $D_I \tilde{x}_0(I; \mu)$ by implicitly differentiating the equation $D_x H(\tilde{x}_0(I; \mu), I; \mu) = 0$. This simple calculation gives

$$D_I \tilde{x}_0(I; \mu) = - \left(D_x^2 H(\tilde{x}_0(I; \mu), I; \mu)\right)^{-1} \left(D_I D_x H(\tilde{x}_0(I; \mu), I; \mu)\right). \qquad (8.17)$$

Equations (8.16) and (8.17) will be useful later on. We remark that invertibility of $D_x^2 H(\tilde{x}_0(I; \mu), I; \mu)^{-1}$ follows from the hyperbolicity of $\tilde{x}_0(I; \mu)$.

8.3.2. The Dynamics on $\mathcal{M}_\epsilon$ near the Resonance. We now want to discuss the dynamics on $\mathcal{M}_\epsilon$ under the perturbed vector fields (8.1) and (8.2). The situation for Hamiltonian and non-Hamiltonian perturbations are very different, so we consider each case individually.

Non-Hamiltonian Perturbations

The perturbed vector field restricted to $\mathcal{M}_\epsilon$ is given by

$$\dot{I} = \epsilon g^I(\tilde{x}_\epsilon(I, \theta, \mu), I, \theta; \mu, \epsilon),$$

$$\dot{\theta} = D_I H(\tilde{x}_\epsilon(I, \theta; \mu), I; \mu) + \epsilon g^\theta(\tilde{x}_\epsilon(I, \theta; \mu), I, \theta; \mu, \epsilon). \tag{8.18}$$

Taylor expanding (8.18) in powers of ϵ gives

$$\dot{I} = \epsilon g^I(\tilde{x}_0(I; \mu), I, \theta; \mu, 0) + \epsilon^2 \left(\langle D_x g^I(\tilde{x}_0(I; \mu), I, \theta; \mu, 0), \tilde{x}_1(I, \theta; \mu) \rangle \right.$$

$$\left. + (D_\epsilon g^I(\tilde{x}_0(I; \mu), I, \theta; \mu, 0) \right) + \mathcal{O}(\epsilon^3),$$

$$\dot{\theta} = D_I H(\tilde{x}_0(I; \mu), I; \mu) + \epsilon \left(\langle D_x \left(D_I H(\tilde{x}_0(I; \mu), I; \mu) \right), \tilde{x}_1(I, \theta; \mu) \rangle \right.$$

$$\left. + g^\theta(\tilde{x}_0(I; \mu), I, \theta; \mu, 0) \right) + \mathcal{O}(\epsilon^2), \tag{8.19}$$

where $\langle \cdot, \cdot \rangle$ represents the usual Euclidean inner product.

We want to study the dynamics of (8.19) near the resonance $I = I^r$. For this purpose we will change variables in order to derive a simpler equation that describes the dynamics in a neighborhood of the resonance. Substituting

$$\begin{aligned} I &= I^r + \sqrt{\epsilon} h, \\ \theta &= \theta, \end{aligned} \tag{8.20}$$

into (8.19), Taylor expanding in I in powers of $\sqrt{\epsilon}$, and rescaling time by letting $\tau = \sqrt{\epsilon} t$ gives the equations

$$h' = g^I + \sqrt{\epsilon} G(h, \theta, \mu) + \mathcal{O}(\epsilon),$$

$$\theta' = \left(\langle D_x(D_I H), D_I \tilde{x}_0 \rangle + D_I^2 H \right) h + \sqrt{\epsilon} F(h, \theta, \mu) + \mathcal{O}(\epsilon), \tag{8.21}$$

where "/" denotes differentiation with respect to the rescaled time τ,

$$G(h, \theta, \mu) = \left(\langle D_x g^I, D_I \tilde{x}_0 \rangle + D_I g^I \right) h,$$

and

$$F(h, \theta, \mu) = \frac{1}{2} (\langle (D_x(D_x D_I H)) D_I \tilde{x}_0, D_I \tilde{x}_0 \rangle + \langle D_x(D_I H), D_I^2 \tilde{x}_0 \rangle$$

$$+ 2 \langle D_x(D_I^2 H), D_I \tilde{x}_0 \rangle + D_I^3 H) h^2 + \langle D_x(D_I H), \tilde{x}_1 \rangle + g^\theta, \tag{8.22}$$

and where all functions are evaluated at $\tilde{x}_0(I; \mu) = \tilde{x}_0(I^r; \mu)$, $I = I^r$, θ, μ, and $\epsilon = 0$. The important advantage gained in localizing the equations near the resonance is that at $\epsilon = 0$, (8.21) is a one-degree-of-freedom (hence, integrable) Hamiltonian system given by

$$h' = g^I = -D_\theta \mathcal{H},$$

$$\theta' = \left(\langle D_x(D_I H), D_I \tilde{x}_0 \rangle + D_I^2 H \right) h = D_h \mathcal{H}, \tag{8.23}$$

where

$$\mathcal{H}(h, \theta) = \left(\langle D_x(D_I H), D_I \tilde{x}_0 \rangle + D_I^2 H \right) \frac{h^2}{2} - \int_{\theta_0}^{\theta} g^I \, d\bar{\theta} \tag{8.24}$$

is the Hamiltonian function. The integrable Hamiltonian structure at leading order is typical near resonances, (e.g., Wiggins [1990]), and is extremely useful for understanding the qualitative (as well as the quantitative) structure of the dynamics near the resonance on $\mathcal{M}_\epsilon$. The $\sqrt{\epsilon}$ dependence in the change of variables given in (8.20) is a consequence of the implicit assumption that

$$\left(\langle D_x(D_I H(\tilde{x}_0(I^r; \mu), I^r; \mu)), D_I \tilde{x}_0 \rangle + D_I^2 H(\tilde{x}_0(I^r; \mu), I^r; \mu) \right) \neq 0.$$

If this is violated, then a scaling with a different fractional power of ϵ is required. (See Wiggins [1990] for details.)

From the point of view of the dynamics on $\mathcal{M}_\epsilon$ we will henceforth only be interested in the dynamics in an $\mathcal{O}(\sqrt{\epsilon})$ neighborhood of the resonance $I = I^r$ and it will be useful to introduce some notation that emphasizes this fact. We will be studying the dynamics in the annulus centered at $I = I^r$ denoted as follows:

$$\mathcal{A}_\epsilon \equiv \{ (x, h, \theta) \mid x = \tilde{x}_\epsilon(I^r + \sqrt{\epsilon} h, \theta, \mu), \ |h| < C \},$$

where $C > 0$ is some constant. The constant C is chosen sufficiently large so that the annulus contains the unperturbed homoclinic orbit. *It is important to note that in the $h - \theta$ coordinates the resonance zone (i.e., the annulus $\mathcal{A}_\epsilon$) is of $\mathcal{O}(1)$ width.* The three-dimensional stable and unstable manifolds of $\mathcal{A}_\epsilon$, denoted $W^s(\mathcal{A}_\epsilon)$ and $W^u(\mathcal{A}_\epsilon)$, respectively, are subsets of $W^s(\mathcal{M}_\epsilon)$ and $W^u(\mathcal{M}_\epsilon)$, respectively, that are obtained by restricting the I values appropriately.

From the point of view of perturbation theory, we will want to compare the dynamics in $\mathcal{A}_\epsilon$ with the unperturbed dynamics in the same region on $\mathcal{M}_\epsilon$ (this will be described more precisely). For this reason we define the "unperturbed annulus" as follows

$$\mathcal{A}_0 \equiv \{ (x, h, \theta) \mid x = \tilde{x}_0(I^r; \mu), \ |h| < C \}$$

with its three-dimensional stable and unstable manifolds $W^s(\mathcal{A}_0)$ and $W^u(\mathcal{A}_0)$ that coincide along a branch.

We make the following assumption on the structure of the integrable Hamiltonian system (8.23).

Assumption 3. For $\mu = \mu_0$, there exists $\theta_c(\mu_0)$ and $\theta_s(\mu_0)$ such that $q_0 = (h, \theta) = (0, \theta_s(\mu_0))$ is a hyperbolic saddle type fixed point of (8.23) and $p_0 = (h, \theta) = (0, \theta_c(\mu_0))$ is a center type fixed point of (8.23). Moreover, q_0 is connected to itself by a homoclinic orbit and p_0 is the only fixed point inside this homoclinic orbit. (See Figure 8.2.)

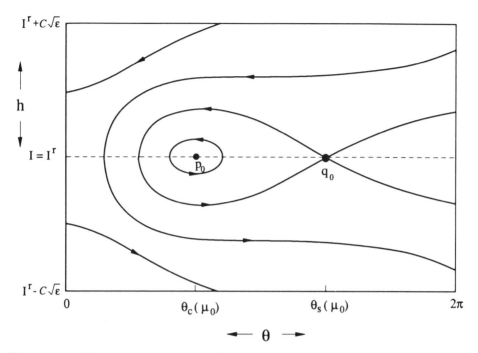

Figure 8.2. The dynamics associated with the leading order Hamiltonian vector field restricted to $\mathcal{M}_\epsilon$ described in **Assumption 3**.

We make the following remarks concerning **Assumption 3** and its consequences.

Remark 1. It is possible for there to be more than one saddle-center pair with a homoclinic connection on the resonance. In this case, our methods can be applied to each pair separately.

Remark 2. Since the matrix associated with the linearization of the vector field about p_0 and q_0 is invertible, these fixed points will typically exist for an open set of parameter values that contains μ_0. A typical way for these fixed points to disappear, as the parameters are varied, is for them to coalesce in a Hamiltonian saddle-node bifurcation. However, if more fixed points and homoclinic (or heteroclinic) orbits exist, then other scenarios are possible. **One should view μ_0 as the set of parameter values for which Assumption 3 holds.**

Remark 3. An equation for the separatrix curve can be easily obtained from the Hamiltonian (8.23) by using the fact that the "energy" of the level set of $\mathcal{H}$ that defines the separatrix is equal to the energy of the saddle point q_0, i.e.,

$$\mathcal{H}(h, \theta, \mu_0) - \mathcal{H}(0, \theta_s(\mu_0), \mu_0) = 0. \tag{8.25}$$

Remark 4. Since (8.23) is an integrable Hamiltonian system, obviously the $\mathcal{O}(\sqrt{\epsilon})$ terms in (8.21) may have a dramatic effect on the phase portrait. In particular, we are interested in the effect on the fixed points q_0 and p_0 and the homoclinic orbit connecting q_0, as well as the dynamics inside the region bounded by this homoclinic orbit. The following two lemmas address these issues.

Lemma 8.3.1. *For ϵ sufficiently small:*

1. *q_0 remains a hyperbolic fixed point of saddle stability type, denoted q_ϵ, for (8.21).*

2. *If*

$$\left(\langle D_x(D_I H), D_\theta \tilde{x}_1 \rangle + D_\theta g^\theta + D_I g^I + \langle D_x g^I, D_I \tilde{x}_0 \rangle \right) < 0 \qquad (8.26)$$

   inside the homoclinic orbit connecting q_0 then p_0 becomes a hyperbolic sink, denoted p_ϵ, for (8.21) and the homoclinic orbit breaks with a branch of the unstable manifold of q_ϵ falling into the sink, p_ϵ, as shown in Figure 8.3.

3. *p_0 is $\mathcal{O}(\sqrt{\epsilon})$ close to p_ϵ and q_0 is $\mathcal{O}(\sqrt{\epsilon})$ close to q_ϵ (where closeness is measured in the $h - \theta$ coordinates).*

Proof: Part 1 follows from the persistence of hyperbolic fixed points. Part 2 uses the fact that the quantity

$$\left(\langle D_x D_I H, D_\theta \tilde{x}_1 \rangle + D_\theta g^\theta + D_I g^I + \langle D_x g^I, D_I \tilde{x}_0 \rangle \right),$$

is just

$$D_\theta F + D_h G,$$

which is the leading order term of the trace of the linearization of (8.21). Then a routine analysis using Bendixson's criteria and some simple phase plane techniques gives the result. (See Wiggins [1990] for details.) Part 3 follows from a simple application of the implicit function theorem.   □

The area enclosed by the homoclinic orbit connecting q_0 in (8.23) is a good approximation to the domain of attraction of the sink p_ϵ. The following lemma makes this more precise.

Lemma 8.3.2. *Suppose condition (8.26) of Lemma 8.3.1. holds and let p denote any point in the region bounded by the unperturbed homoclinic orbit that is an $\mathcal{O}(1)$ distance (in the $h - \theta$ coordinates) from the homoclinic orbit. Then, for ϵ sufficiently small, under the perturbed dynamics (i.e., (8.21)) the trajectory through p approaches p_ϵ asymptotically as $\tau \to \infty$.*

Proof: Under condition (8.26) of Lemma 8.3.1, the stable manifold of q_ϵ forms the boundary of the basin of attraction for p_ϵ. A standard planar Melnikov analysis of (8.21) shows that the stable and unstable manifolds of q_ϵ split by an $\mathcal{O}(\sqrt{\epsilon})$ amount (in the $h - \theta$ coordinates) in a tubular neighborhood of the unperturbed

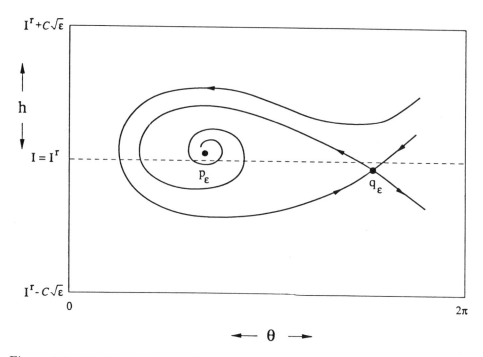

Figure 8.3. The dynamics near the resonance of the vector field restricted to $\mathcal{M}_\epsilon$ for the full vector field.

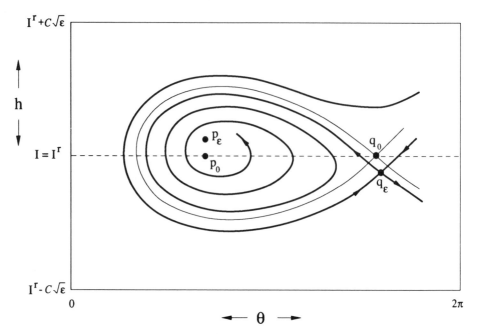

Figure 8.4. The basin of attraction of p_ϵ and its comparison with the unperturbed structure (for the "slow time " system) near the resonance.

homoclinic orbit *excluding a small fixed neighborhood of q_0*, hence the result follows. (See Figure 8.4.)   □

Henceforth, for the case of non-Hamiltonian perturbations (8.1), we will assume that Assumptions 1 to 3 hold as well as (8.26).

Hamiltonian Perturbations

We now want to describe the dynamics on $\mathcal{A}_\epsilon$ under Hamiltonian perturbations. We first prove a technical lemma which will not be stated in its full generality, rather in a form adapted to our problem. Throughout the lemma d will refer to exterior differentiation and $i_X\omega$ will denote the interior product of the vectorfield X with the form ω.

Lemma 8.3.3. *Let $(\mathcal{P},\omega)$ be a four-dimensional symplectic C^r manifold with (locally defined) coordinates $x = (x_1, x_2) \in \mathbb{R}^2$ and $z = (z_1, z_2) \in \mathbb{R}^2$, and with the symplectic form $\omega = dx_1 \wedge dx_2 + dz_1 \wedge dz_2$. Consider a one parameter family of two dimensional C^k submanifolds $\mathcal{A}_\epsilon \subset \mathcal{P}$, with $1 \le k \le r$ and $\epsilon \in [0, \epsilon_0]$, of the form*

$$\mathcal{A}_\epsilon = \{(x, z) \in \mathcal{P}\,|\,x = f(z_2) + \epsilon g(z; \epsilon)\},$$

where $f = (f_1, f_2)$ and $g = (g_1, g_2)$ are (locally defined) C^{k+1} functions with $f_j, g_j :$ $\mathbb{R}^2 \to \mathbb{R}$, $j = 1, 2$. We further assume that H is a C^2 function on $\mathcal{P}$, and for any $\epsilon \in [0, \epsilon_0]$, $\mathcal{A}_\epsilon$ is an invariant manifold for the Hamiltonian vectorfield $X_H : \mathcal{P} \to$ $T\mathcal{P}$ defined by

$$i_{X_H}\omega = dH.$$

Then, for ϵ sufficiently small:

1. *$(\mathcal{A}_\epsilon, \bar{\omega}_\epsilon)$ is a two dimensional symplectic C^k manifold with $\bar{\omega}_\epsilon = \omega|\mathcal{A}_\epsilon$.*

2. *$X_\epsilon = X_H|\mathcal{A}_\epsilon$ is a Hamiltonian vectorfield on $\mathcal{A}_\epsilon$ with Hamiltonian $\mathcal{H}_\epsilon = H|\mathcal{A}_\epsilon$, i.e.*

$$i_{X_\epsilon}\bar{\omega}_\epsilon = d\mathcal{H}_\epsilon.$$

Proof: We first show that $\bar{\omega}_\epsilon$ defines a symplectic structure on $\mathcal{A}_\epsilon$. On $\mathcal{A}_\epsilon$ we have

$$dx_j = \epsilon D_{z_1}g_j dz_1 + D_{z_2}(f_j + \epsilon g_j)dz_2, \quad j = 1, 2,$$

implying

$$\bar{\omega}_\epsilon = \epsilon\big(D_{z_1}g_1 D_{z_2}(f_2 + \epsilon g_2) - D_{z_2}(f_1 + \epsilon g_1)D_{z_1}g_2\big)dz_1 \wedge dz_2$$

$$+dz_1 \wedge dz_2 \qquad (8.27)$$

$$= \big(1 + \mathcal{O}(\epsilon)\big)dz_1 \wedge dz_2,$$

from which we conclude that, for ϵ sufficiently small, $\bar{\omega}_\epsilon$ is a nondegenerate 2-form on $\mathcal{A}_\epsilon$. Let $e_\epsilon = (f + \epsilon g, Id_2) : \mathbb{R}^2 \to \mathcal{P}$ be the embedding of $\mathcal{A}_\epsilon$. Then

$$d\bar{\omega}_\epsilon = d(e_\epsilon^*\omega) = e_\epsilon^*d\omega = 0, \qquad (8.28)$$

which shows that $\bar{\omega}_\epsilon$ is closed (e_ϵ^* denotes the pull-back of e_ϵ). But (8.27) and (8.28) together imply statement 1 of the lemma.

To prove 2 we consider an arbitrary $p_\epsilon \in e_\epsilon^{-1}(\mathcal{A}_\epsilon)$ and $u \in T_p\mathbb{R}^2$. We can write

$$i_{X_\epsilon}\bar{\omega}_\epsilon[p_\epsilon](u) = e^*\omega[p_\epsilon]\big(e_\epsilon * X_H(p_\epsilon), u\big)$$

$$= \omega[e_\epsilon(p_\epsilon)]\big(de_\epsilon \, de_\epsilon^{-1}X_H\big(e_\epsilon(p_\epsilon)\big), de_\epsilon \, u\big)$$

$$= \omega[e_\epsilon(p_\epsilon)]\big(X_H\big(e_\epsilon(p_\epsilon)\big), de_\epsilon \, u\big) \qquad (8.29)$$

$$= dH[e_\epsilon(p_\epsilon)](de_\epsilon \, u) = d(e_\epsilon^*H)[p_\epsilon](u)$$

$$= d\mathcal{H}_\epsilon[p_\epsilon](u),$$

which proves 2 of the lemma. $\square$

We make the following remarks concerning this result.

Remark 1 The dimensional restriction on z in Lemma 8.3.3 is not essential but simplifies the proof of 1. Furthermore, f can be a function of z_1 only.

Remark 2 Lemma 8.3.3 is easy, but not trivial, because it is obviously not true for *arbitrary* invariant manifolds of the Hamiltonian dynamics generated by H, e.g. for stable and unstable manifolds of equilibria. These latter manifolds cannot be viewed as graphs over two variables which are canonically conjugate, as it is readily seen from the structure of the stable and unstable subspaces of the equilibrium in the linearized problem.

We next apply this lemma to (8.2) restricted to $\mathcal{A}_\epsilon$.

Proposition 8.3.1. *Consider (8.2) restricted to $\mathcal{A}_\epsilon$. For ϵ small enough, the restricted dynamics is Hamiltonian with Hamiltonian*

$$\mathcal{H}_\epsilon(I,\theta;\mu)|_{I=I^r+\sqrt{\epsilon}h} = H(\tilde{x}_0(I^r;\mu),I^r;\mu)$$

$$+\epsilon[\tfrac{1}{2}m(I^r;\mu)h^2 + H_p(\tilde{x}_0(I^r;\mu),I^r,\theta;\mu)] + \mathcal{O}(\epsilon^{\frac{3}{2}}), \tag{8.30}$$

with

$$m(I^r;\mu) = D_I^2 H(\tilde{x}_0(I^r;\mu),I^r;\mu)$$

and with the restricted symplectic form

$$\bar{\omega}_\epsilon = (1+\mathcal{O}(\epsilon))d\theta \wedge dI. \tag{8.31}$$

Proof: We can directly apply Lemma 8.3.3 with $z_1 = \theta$, $z_2 = I$ and $f = \tilde{x}_0(I^r;\mu)$ to obtain (8.31). We outline the steps. First, recall that the full Hamiltonian is given by

$$H(x,I,\theta;\mu,\epsilon) = H(x,I;\mu) + \epsilon H_1(x,I,\theta;\mu,\epsilon).$$

| Restrict the Full Hamiltonian to $\mathcal{A}_\epsilon$ and Taylor Expand in ϵ. |

Doing this, we obtain

$$\begin{aligned}\mathcal{H}_\epsilon(I,\theta;\mu) &\equiv H(\tilde{x}_\epsilon(I,\theta;\mu),I,\theta;\mu,\epsilon), & (8.32)\\ &= H(\tilde{x}_0(I;\mu),I;\mu) + \epsilon H_p(\tilde{x}_0(I;\mu),I,\theta;\mu) + \mathcal{O}(\epsilon^2), & (8.33)\end{aligned}$$

where

$$H_p(\tilde{x}_0(I;\mu),I,\theta;\mu) \equiv H_1(\tilde{x}_0(I;\mu),I,\theta;\mu,0) \tag{8.34}$$

and we have used Assumption 1 (i.e., the fact that $\tilde{x}_0(I;\mu)$ are equilibria for the x-component of the unperturbed problem).

| Localize the Hamiltonian Near the Resonance. |

We make the following change of variables in (8.33):

$$I \mapsto I^r + \sqrt{\epsilon}h,$$
$$\theta \mapsto \theta,$$

and Taylor expand in $\sqrt{\epsilon}$, using Assumption 2 (i.e., a resonant circle of fixed points at $I = I^r$) to obtain (8.30). The result then follows from Proposition 8.3.1.   □

Note that the change of variables

$$I \mapsto I^r + \sqrt{\epsilon}h,$$
$$\theta \mapsto \theta,$$

is not canonical since

$$(1 + \mathcal{O}(\epsilon))d\theta \wedge dI = \sqrt{\epsilon}(1 + \mathcal{O}(\epsilon))d\theta \wedge dh, \tag{8.35}$$

but this also shows that if we introduce the new restricted Hamiltonian

$$\bar{\mathcal{H}}_\epsilon(h, \theta; \mu) = \tfrac{1}{\sqrt{\epsilon}(1+\mathcal{O}(\epsilon))} \left(\mathcal{H}_\epsilon(I(h; \epsilon), \theta; \mu) - H(\tilde{x}_0(I^r; \mu), I^r; \mu) \right)$$
$$= \sqrt{\epsilon}\mathcal{H}(h, \theta; \mu) + \mathcal{O}(\epsilon) \tag{8.36}$$

with

$$\mathcal{H}(h, \theta; \mu) = \frac{1}{2}m(I^r; \mu)h^2 + H_p(\tilde{x}_0(I^r; \mu), I^r, \theta; \mu), \tag{8.37}$$

(the $\mathcal{O}(\epsilon)$ term is just the same that appeared in (8.35)) then we can derive the leading order Hamiltonian dynamics of $\mathcal{A}_\epsilon$ from (8.36) in the canonical way (i.e. using the symplectic form $d\theta \wedge dh$) to obtain

$$\dot{h} = -\sqrt{\epsilon}D_\theta\mathcal{H}(h, \theta; \mu) + \mathcal{O}(\epsilon),$$
$$\dot{\theta} = \sqrt{\epsilon}D_h\mathcal{H}(h, \theta; \mu) + \mathcal{O}(\epsilon), \tag{8.38}$$

which we call the *restricted system*. Rescaling the time in (8.38) by letting $\tau = \sqrt{\epsilon}t$, denoting $\frac{d}{d\tau}$ by $'$ and setting $\epsilon = 0$, we arrive at the equations

$$h' = -D_\theta\mathcal{H}(h, \theta; \mu),$$
$$\theta' = D_\theta\mathcal{H}(h, \theta; \mu), \tag{8.39}$$

with Hamiltonian $\mathcal{H}(h, \theta; \mu)$, which we call the *reduced system*. We will consider the reduced systems as defined on the annulus

$$A = [-h_0, h_0] \times S^1,$$

where h_0 is chosen large enough to contain the orbits of interest. We note that for ϵ sufficiently small $\mathcal{A}_\epsilon$ and A are diffeomorphic. Notice that (8.39) describes a simple one-degree-of-freedom potential problem: $m(I^r; \mu)$ can be thought of as a mass of the particle and $H_p(\tilde{x}_0(I^r; \mu), I^r, \theta; \mu)$ can be viewed as the potential of the forces acting on the particle. Measured in the (h, θ) coordinates, the orbits of this potential problem perturb by an amount of $\mathcal{O}(\sqrt{\epsilon})$ into the orbits of (8.38), so just by analysing this simple system we can obtain some information about the orbit structure of the restricted problem (8.38). However, we have to be careful at this point for two reasons:

1. Orbits of the reduced system might leave the annulus in finite time or might perturb to orbits of (8.38) that leave the annulus. In either case, the part of the orbit outside the annulus is meaningless for us.

2. Certain orbits of the reduced system (e.g. orbits homoclinic to singular equilibria, heteroclinic orbits) might not perturb smoothly into nearby orbits of (8.38).

To discuss and formulate these problems we introduce the following definition:

Definition 8.3.1. *We say that an orbit γ of some Hamiltonian system defined on the annulus A is an internal orbit if both of the following are satisfied:*

1. *γ is either a periodic orbit or an orbit homoclinic to a hyperbolic fixed point.*

2. *γ is bounded away from the boundary of the annulus.*

Since for ϵ sufficiently small $\mathcal{A}_\epsilon$ is diffeomorphic to A we will interchangeably speak of internal orbits in $\mathcal{A}_\epsilon$ and A.

Remark: γ is considered to be a homoclinic orbit if it connects points which have equal h coordinates and equal θ coordinates mod 2π. In specific applications, one can also allow γ to be a heteroclinic orbit if it is structurally stable with respect to the class of perturbations considered.

It is clear from Definition 8.3.1 that internal orbits do not intersect and are not asymptotic to the boundary of the annulus, and this property is obviously preserved for sufficiently small perturbations. Furthermore, in case of small perturbation of the Hamiltonian system, internal orbits deform smoothly into nearby orbits of the perturbed system. The reason why we have also excluded nonsingular equilibria from the definition of internal orbits will become apparent shortly.

8.3.3. The Fibering of $W^s(\mathcal{A}_\epsilon)$ and $W^u(\mathcal{A}_\epsilon)$: The Singular Perturbation Nature. Theorem 8.3.1 provides a description of the asymptotic behavior of orbits in $W^s(\mathcal{A}_\epsilon)$ and $W^u(\mathcal{A}_\epsilon)$, and Lemmas 8.3.1 and 8.3.2 describe the dynamics on $\mathcal{A}_\epsilon$. Now we want to "tie together" these results. In particular, we want to characterize orbits in $W^s(\mathcal{A}_\epsilon)$ and $W^u(\mathcal{A}_\epsilon)$ in terms of the orbits to which they asymptote in $\mathcal{A}_\epsilon$. We will see that this is a delicate problem in *singular perturbation theory* since the dynamics on $\mathcal{A}_\epsilon$ are "slow" compared to the "fast" dynamics transverse to $\mathcal{A}_\epsilon$.

Indeed, the dynamics near the resonance as described earlier are created entirely by the perturbation. For these issues the (x, h, θ) coordinates will be most appropriate.

More insight into these questions can be obtained by directly examining the equations of motion. As described in the previous section, the dynamics on $\mathcal{A}_\epsilon$ can be studied by restricting the vector field to $\mathcal{A}_\epsilon$ and introducing coordinates localized near the resonance as in (8.20). We rewrite these equations below.

Non-Hamiltonian Perturbations

$$\dot{h} = \sqrt{\epsilon}g^I + \epsilon G(h, \theta, \mu) + \mathcal{O}(\epsilon^{3/2}),$$

$$\dot{\theta} = \sqrt{\epsilon}\left(\langle D_x(D_I H), D_I \tilde{x}_0 \rangle + D_I^2 H\right) h + \epsilon F(h, \theta, \mu) + \mathcal{O}(\epsilon^{3/2}). \tag{8.40}$$

Note that, for $\epsilon = 0$, (8.40) reduces to

$$\begin{aligned} \dot{h} &= 0, \\ \dot{\theta} &= 0. \end{aligned} \tag{8.41}$$

Of course, the dynamics in the full phase space are described by

$$\dot{x} = JD_x H(x, I^r, \mu) + \sqrt{\epsilon}D_I(JD_x H(x, I^r))h + \tfrac{\epsilon}{2}D_I^2(JD_x H(x, I^r, \mu))h^2$$

$$+\epsilon g^x(x, I^r, \theta, \mu, 0) + \mathcal{O}(\epsilon^{3/2}),$$

$$\dot{h} = \sqrt{\epsilon}g^I(x, I^r, \theta, \mu, 0) + \epsilon D_I g^I(x, I^r, \theta, \mu, 0)h + \mathcal{O}(\epsilon^{3/2}), \tag{8.42}$$

$$\dot{\theta} = D_I H(x, I^r, \mu) + \sqrt{\epsilon}D_I^2 H(x, I^r, \mu)h + \tfrac{\epsilon}{2}D_I^3 H(x, I^r, \mu)h^2$$

$$+\epsilon g^\theta(x, I^r, \theta, \mu, 0) + \mathcal{O}(\epsilon^{3/2}).$$

For $\epsilon = 0$, (8.42) reduces to

$$\dot{x} = JD_x H(x, I^r, \mu),$$

$$\dot{h} = 0, \tag{8.43}$$

$$\dot{\theta} = D_I H(x, I^r, \mu).$$

Hamiltonian Perturbations

$$\begin{aligned} \dot{h} &= -\sqrt{\epsilon}D_\theta \mathcal{H}(h, \theta; \mu) + \mathcal{O}(\epsilon), \\ \dot{\theta} &= \sqrt{\epsilon}D_h \mathcal{H}(h, \theta; \mu) + \mathcal{O}(\epsilon). \end{aligned} \tag{8.44}$$

At $\epsilon = 0$, (8.44) reduces to

$$\dot{h} = 0,$$
$$\dot{\theta} = 0. \tag{8.45}$$

The dynamics in the full phase space are described by

$$\dot{x} = JD_x H(x, I^r; \mu) + \sqrt{\epsilon} JD_x D_I H(x, I^r; \mu)h + \mathcal{O}(\epsilon),$$

$$\dot{h} = -\sqrt{\epsilon} D_\theta H_p(x, I^r, \theta; \mu) + \mathcal{O}(\epsilon^{\frac{3}{2}}), \tag{8.46}$$

$$\dot{\theta} = D_I H(x, I^r; \mu) + \sqrt{\epsilon} D_I^2 H(x, I^r; \mu)h + \mathcal{O}(\epsilon).$$

At $\epsilon = 0$, these equations reduce to

$$\dot{x} = JD_x H(x, I^r; \mu),$$

$$\dot{h} = 0, \tag{8.47}$$

$$\dot{\theta} = D_I H(x, I^r; \mu).$$

There are several features that we want to point out concerning the above sets of equations.

1. From (8.41) and (8.45), we see that at $\epsilon = 0$ the neighborhood of the resonance on $\mathcal{A}$ in the variables scaled as in (8.20) consists entirely of fixed points. We can think of the change of variables in (8.20) as "blowing up" the circle of fixed points into an annulus of fixed points centered at $I = I^r$.

2. For ϵ small, but nonzero, we see that, roughly speaking, the character of the dynamics in the x variables is not altered much under the influence of the perturbation. Theorem 8.3.1 makes this more precise. However, the dynamics on the annulus is radically different. Indeed, for $\epsilon = 0$ there are no dynamics on the annulus (it consists entirely of fixed points), whereas for ϵ small, the typical resonance structure is created as was revealed through a study of (8.40) under the rescaled time or, *slow time*, $\tau = \sqrt{\epsilon}t$.

As we mentioned earlier, we want to relate the asymptotic behavior of trajectories in the stable and unstable manifolds of the annulus to trajectories in the annulus for ϵ small, but nonzero. From equations (8.40) to (8.47) as well as the discussion following these equations, it should be clear that this is a singular perturbation problem; however, since we will require infinite-time results, classical singular perturbation approaches will be of limited use. Rather, we will see that the problem is most naturally addressed from the geometrical, dynamical systems viewpoint that utilizes the *fibering* of the stable and unstable manifolds by submanifolds consisting of initial conditions of trajectories that have the same "asymptotic phase".

We will set the stage for this by noting that one can interpret these unperturbed heteroclinic orbits connecting the fixed points in the annulus in a slightly different way. Clearly, all points on a heteroclinic orbit approach the same fixed point in the annulus asymptotically in forward time as well as the same fixed point in the annulus asymptotically in backward time. (Of course, the forward and backward time limit points are, in general, different.) Thus, we see that $W^s(\mathcal{A}_0)$ and $W^u(\mathcal{A}_0)$ can be viewed as the union of a two-parameter family of curves (the two parameters are h and θ) that have the properties that points on the curves asymptotically approach the same orbit (which is just a fixed point) in $\mathcal{A}_0$. We say that $W^s(\mathcal{A}_0)$ and $W^u(\mathcal{A}_0)$ are *fibered* by an invariant family of curves made up of initial conditions of trajectories that asymptote to the same orbit in $\mathcal{A}_0$. In the case of the perturbed problem, if we could prove that these fibers, as well as the dynamical interpretation that all points on the fibers have the same "asymptotic phase", were persistent under perturbation, then we would have a tool for relating the dynamics in $W^s(\mathcal{A}_\epsilon)$ and $W^u(\mathcal{A}_\epsilon)$ to the dynamics in $\mathcal{A}_\epsilon$. This fibering of the stable manifold allows us to view some aspects of a singular perturbation problem as a problem in regular perturbation theory.

Fiberings of stable (and unstable) manifolds of this form have been studied in the context of dynamical systems theory for many years; however, they have not been used much in applications. Fenichel was among the first to apply such ideas to problems arising in singular perturbation theory; in particular, to problems of the form of (8.42). Recall again the geometrical structure of this problem. At $\epsilon = 0$ the annulus $\mathcal{A}_0$ consists entirely of fixed points (under the "fast time" t). By hypothesis, each fixed point is connected to another (in general) fixed point by a heteroclinic connection. Thus the three-dimensional stable, and unstable, manifolds of the annulus are fibered by a two-parameter family (the parameters label a point on the annulus) of one-dimensional fibers; clearly, the fibers at $\epsilon = 0$ are just the unperturbed heteroclinic orbits. Fenichel has proven that the fibers perturb smoothly in $\sqrt{\epsilon}$ and are $\mathcal{O}(\sqrt{\epsilon})$ close to the unperturbed fibers. This is stated more precisely in the following theorem that applies for both (8.42) and (8.46).

Theorem 8.3.2. *There exists $\delta_0 > 0$ and $\epsilon_0 > 0$ such that given any point $(\bar{h}, \bar{\theta}) \in \mathcal{A}_\epsilon$, there exists a family of one-dimensional curves, called the **stable fibers**, that can be represented as graphs as follows*

$$x_2 = x_2(x_1; \bar{h}, \bar{\theta}, \mu, \sqrt{\epsilon}),$$

$$h = h(x_1; \bar{h}, \bar{\theta}, \mu, \sqrt{\epsilon}), \tag{8.48}$$

$$\theta = \theta(x_1; \bar{h}, \bar{\theta}, \mu, \sqrt{\epsilon}),$$

*where $x \equiv (x_1, x_2)$. The point $(\bar{h}, \bar{\theta})$ is referred to as the **base point** of the fiber. These graphs are defined for any $0 < \delta \leq \delta_0, 0 < \epsilon \leq \epsilon_0$ with $|x_1| \leq \delta$ and $\mu \in R^p$. Moreover, these curves have the following properties:*

1. *They are C^r in x_1 and C^{r-1} in $(\bar{h}, \bar{\theta}, \mu, \sqrt{\epsilon})$.*

2.

$$x_2(\tilde{x}_{1\epsilon}(I^r + \sqrt{\epsilon}\bar{h}, \bar{\theta}, \mu); \bar{h}, \bar{\theta}, \mu, \sqrt{\epsilon}) = \tilde{x}_{2\epsilon}(I^r + \sqrt{\epsilon}\bar{h}, \bar{\theta}, \mu),$$

$$h(\tilde{x}_{1\epsilon}(I^r + \sqrt{\epsilon}\bar{h}, \bar{\theta}, \mu); \bar{h}, \bar{\theta}, \mu, \sqrt{\epsilon}) = \bar{h}, \tag{8.49}$$

$$\theta(\tilde{x}_{1\epsilon}(I^r + \sqrt{\epsilon}\bar{h}, \bar{\theta}, \mu); \bar{h}, \bar{\theta}, \mu, \sqrt{\epsilon}) = \bar{\theta},$$

where, recall, $\mathcal{A}_\epsilon$, with the I values suitably restricted, is the graph of $\tilde{x}_\epsilon$ $(I, \theta, \mu) \equiv (\tilde{x}_{1\epsilon}(I, \theta, \mu), \tilde{x}_{2\epsilon}(I, \theta, \mu))$ (cf. Theorem 8.3.1).

3. $W^s_{loc}(\mathcal{A}_\epsilon) = \displaystyle\bigcup_{(\bar{h}, \bar{\theta}) \in \mathcal{A}_\epsilon} \left[x_2(x_1; \bar{h}, \bar{\theta}, \mu, \sqrt{\epsilon}) \right].$

4. Let $(\bar{h}(t), \bar{\theta}(t))$ be a trajectory in $\mathcal{A}_\epsilon$ satisfying $(\bar{h}(0), \bar{\theta}(0)) = (\bar{h}, \bar{\theta})$ and let $(x_1(t), x_2(t), h(t), \theta(t))$ be a trajectory in $W^s_{loc}(\mathcal{A}_\epsilon)$ satisfying

$$x_2(0) = x_2(x_1(0); \bar{h}, \bar{\theta}, \mu, \sqrt{\epsilon}),$$

$$h(0) = h(x_1(0); \bar{h}, \bar{\theta}, \mu, \sqrt{\epsilon}), \tag{8.50}$$

$$\theta(0) = \theta(x_1(0); \bar{h}, \bar{\theta}, \mu, \sqrt{\epsilon}),$$

i.e., the trajectory starts on the fiber with base point $(\bar{h}, \bar{\theta})$, then

$$\left| (x(t), h(t), \theta(t)) - \left(\tilde{x}_\epsilon(I^r + \sqrt{\epsilon}\bar{h}(t), \bar{\theta}(t), \mu), \bar{h}(t), \bar{\theta}(t) \right) \right| < C e^{-\lambda t},$$

for all $t > 0$ and for some C, $\lambda > 0$ as long as $(\bar{h}(t), \bar{\theta}(t)) \in \mathcal{A}_\epsilon$. In other words, trajectories starting on a stable fiber asymptotically approach the trajectory in $\mathcal{A}_\epsilon$ that starts on the base point of the same fiber, as long as the trajectory through this base point remains in $\mathcal{A}_\epsilon$.

5. The family of fibers form an invariant family in the sense that fibers map to fibers under the time t flow map. Analytically, this is expressed as follows. Suppose $(x_1(t), x_2(t), h(t), \theta(t))$ is a trajectory satisfying

$$x_2(0) = x_2(x_1(0); \bar{h}, \bar{\theta}, \mu, \sqrt{\epsilon}),$$

$$h(0) = h(x_1(0); \bar{h}, \bar{\theta}, \mu, \sqrt{\epsilon}), \tag{8.51}$$

$$\theta(0) = \theta(x_1(0); \bar{h}, \bar{\theta}, \mu, \sqrt{\epsilon}),$$

then

$$x_2(t) = x_2(x_1(t); \bar{h}(t), \bar{\theta}(t), \mu, \sqrt{\epsilon}),$$

$$h(t) = h(x_1(t); \bar{h}(t), \bar{\theta}(t), \mu, \sqrt{\epsilon}), \tag{8.52}$$

$$\theta(t) = \theta(x_1(t); \bar{h}(t), \bar{\theta}(t), \mu, \sqrt{\epsilon}).$$

6. At $\epsilon = 0$ the unperturbed fibers correspond to the unperturbed heteroclinic orbits. Hence, the perturbed and unperturbed fibers are $C^r \sqrt{\epsilon}$ -close.

Proof: This follows immediately from the results of Fenichel [1974], [1979]. $\square$

We make the following remarks concerning this theorem.

Remark 1. An identical result follows for the fibering of $W^u(\mathcal{A}_\epsilon)$.

Remark 2. Quasilinear partial differential equations whose solutions are the fibers can be derived. These equations are analogous to those given following Theorem 3.1. We will not require these for our calculations; however, the reader can find these equations in Fenichel [1979].

Finally, we want to state a result that will be important in the next section.

Proposition 8.3.2. $W_{loc}^u(p_\epsilon)$ is C^r ϵ-close to $W_{loc}^u(p_0)$.

Proof: This result follows from a slight, but straightforward, modification of the usual unstable manifold theorem. (See Kovačič [1989] for details.) A modification of the usual result is required since p_0 is not hyperbolic. $\square$

At this stage it is appropriate to warn the reader that "closeness" is a concept that depends on the specific coordinate system under consideration. In particular, in this paper we are considering two coordinate systems: the $x - I - \theta$ coordinate system and the $x - h - \theta$ coordinate system. *Points that are $\mathcal{O}(\epsilon)$ close in the $x-I-\theta$ coordinates are $\mathcal{O}(\sqrt{\epsilon})$ close in the $x-h-\theta$ coordinates.* Thus, Proposition 8.3.2 is a statement about the closeness of $W_{loc}^u(p_\epsilon)$ and $W_{loc}^u(p_0)$ in the $x - I - \theta$ coordinates. These issues will play an important role in the next section.

8.4. Melnikov's Method for Detecting Orbits Homoclinic and Heteroclinic to Invariant Sets in $\mathcal{A}_\epsilon$

Recall the general set-up for the development of "Melnikov type" methods discussed earlier.

1. Construct "homoclinic coordinates" from the unperturbed integrable structure in order to describe the manner in which the homoclinic structures perturb with respect to the unperturbed structures.

2. Using these homoclinic coordinates, set up the distance measurement between the manifolds of interest.

3. Appeal to general results–persistence of normally hyperbolic invariant manifolds, perturbations of fibers, persistence of transversal intersections, etc. – in order to justify the analysis of certain perturbed structures with respect to the homoclinic coordinates.

4. Apply "Melnikov's trick" in order to develop a perturbative, computable expression for the measure of the distance between the perturbed manifolds of interest.

This program can be carried out in the same manner for this case. We first describe the homoclinic coordinates. Recall that the three-dimensional homoclinic manifold can be described by the following equation:

$$W^s(\mathcal{A}_0) \cap W^u(\mathcal{A}_0) = \{(x, I, \theta) \mid H(x, I; \mu) - H(\tilde{x}_0(I; \mu), I; \mu) = 0\},$$

where I is restricted to the domain of definition of $\mathcal{A}_0$. Moreover, it can be parametrized using the unperturbed homoclinic orbits as follows:

$$(t_0, I, \theta_0) \quad \mapsto \quad \left(x^h(t - t_0, I), I, \theta(t - t_0, I, \theta_0) = \int\limits_{-\infty}^{t-t_0} D_I H(x^h(s, I), I, \mu) ds + \theta_0 \right)$$

$$= \quad q_0(t - t_0, I, \theta_0; \mu). \tag{8.53}$$

Note that the point $(t_0, I, \theta_0) = (+\infty, I, \theta_0)$ corresponds to a point on $\mathcal{A}_0$. We will denote a general point on the homoclinic manifold by $p = (t_0, I, \theta_0)$. A vector normal to the homoclinic manifold at the point p is given by

$$n(p) = (D_x H(x, I; \mu), D_I (H(x, I; \mu) - H(\tilde{x}_0(I; \mu), I; \mu)), 0).$$

Now, in the set-up developed thus far for orbits homoclinic to resonances, there are a variety of different invariant sets for which we can consider homoclinic or heteroclinic connections. We begin by considering the intersection of the three-dimensional stable and unstable manifolds of $\mathcal{A}_\epsilon$.

$$\boxed{W^s(\mathcal{A}_\epsilon) \cap W^u(\mathcal{A}_\epsilon)}$$

The transversal intersection of these two three-dimensional sets would be two-dimensional and represent a one-parameter set of trajectories that are backward and forward asymptotic to $\mathcal{A}_\epsilon$. In Wiggins [1988a] and Kovačič and Wiggins [1992] it is shown that an $\mathcal{O}(\epsilon)$ approximation to the distance between $W^s(\mathcal{A}_\epsilon)$ and $W^u(\mathcal{A}_\epsilon)$ is given by

$$M(t_0, I, \theta_0; \mu) = \int_{-\infty}^{+\infty} \langle n, (g^x, g^I, g^\theta) \rangle (q_0(t - t_0, I, \theta_0; \mu), \mu, 0) \, dt, \tag{8.54}$$

which we refer to as the *Melnikov function* because it is derived, as discussed at the beginning of this section, in exactly the same spirit as Melnikov's original arguments. Note that if we make the change of variables $t \to t + t_0$ in (8.54) then t_0 disappears from the Melnikov function. Henceforth, we assume this has been done and omit t_0 from the arguments of the Melnikov function. This has an important geometrical meaning that we now describe.

If the three-dimensional $W^u(\mathcal{A}_\epsilon)$ and the three-dimensional $W^s(\mathcal{A}_\epsilon)$ intersect at a single point, then, by uniqueness of solutions, they must coincide along a one-dimensional orbit. This is the reason why t_0 can be eliminated from the Melnikov function, since t_0 parametrizes points along a trajectory. Hence, the measure of distance between the perturbed manifolds can be made at any point along the unperturbed trajectory. Henceforth, we will write this Melnikov function as

$$M(I, \theta_0; \mu) = \int_{-\infty}^{+\infty} \langle n, (g^x, g^I, g^\theta) \rangle \left(q_0\left(t, I, \theta_0; \mu\right), \mu, 0 \right) dt. \qquad (8.55)$$

In particular, it is shown through an application of the implicit function theorem that simple zeros of this function correspond to transverse intersections of $W^s(\mathcal{A}_\epsilon)$ and $W^u(\mathcal{A}_\epsilon)$. Trajectories in $W^s(\mathcal{A}_\epsilon) \cap W^u(\mathcal{A}_\epsilon)$ approach $\mathcal{A}_\epsilon$ in forward and backward time, however, without additional analysis, we cannot say anything more. The tool that enables us to learn more about the forward and backward time behavior of these orbits is the perturbation theory for foliations of stable and unstable manifolds.

$$\boxed{W^s(\mathcal{A}_\epsilon) \cap W^u(p_\epsilon)}$$

These three and one-dimensional invariant sets cannot intersect transversely, since the intersection must be a one-dimensional trajectory. Moreover, we do not know if p_ϵ is homoclinically connected to itself or heteroclinically connected to another invariant set in $\mathcal{A}_\epsilon$. This requires an additional analysis that will be performed after the conditions under which $W^u(p_\epsilon)$ intersects $W^s(\mathcal{A}_\epsilon)$ are determined.

First note that the trajectory in $W^u(\mathcal{A}_0) \cap W^s(\mathcal{A}_0)$ that is backwards time asymptotic to p_0 is given by

$$q_0(t, I^r, \theta_c(\mu_0), \mu_0) = \left(x^h(t, I^r), I^r, \theta(t, I^r) \right) \qquad (8.56)$$

$$= \int_{-\infty}^{t} D_I H(x^h(s, I^r), I^r, \mu_0) ds + \theta_c(\mu_0) \ ).$$

Evaluating the Melnikov function (8.55) on this particular trajectory gives

$$M(I^r, \theta_c(\mu_0); \mu_0) = \int_{-\infty}^{+\infty} \langle n, (g^x, g^I, g^\theta) \rangle \left(q_0(t, I^r, \theta_c(\mu_0)), \mu_0, 0 \right) dt. \qquad (8.57)$$

It is shown in Kovačič and Wiggins [1992] that this is the leading order term (in an expansion in ϵ) for the distance between $W^s(\mathcal{A}_\epsilon)$ and $W^u(p_\epsilon)$. More precisely, we have the following theorem.

Theorem 8.4.1. *Suppose that at* $\mu = \mu_0$

1. $M(I^r, \theta_c(\mu_0), \mu_0) = 0,$

2. $\frac{d}{d\mu}M(I^r, \theta_c(\mu_0), \mu_0)$ *has rank 1,*

then $W^u(p_\epsilon) \subset W^s(\mathcal{A}_\epsilon)$.

Note that the Melnikov function is only a function of the parameter μ_0. This is not unexpected since we would expect that the one-dimensional $W^u(p_\epsilon)$ and the three-dimensional $W^s(\mathcal{A}_\epsilon)$ would generically intersect in a one-parameter family of vector fields, i.e., it is a codimension one phenomenon.

$$W^u(p_\epsilon) \cap W^s(p_\epsilon): \text{ A Homoclinic Orbit to } p_\epsilon$$

Our arguments in this section will apply to the rescaled equations (8.40) and (8.42). *In particular, all measures of "closeness" will be made with respect to the* $x - h - \theta$ *coordinates.* Suppose we have shown that $W^u(p_\epsilon) \subset W^s(\mathcal{A}_\epsilon)$, then it follows from Proposition 8.3.2 that there exist points

$$
\begin{aligned}
\hat{p}_\epsilon &\equiv \partial U^\delta \cap W^u_{loc}(p_\epsilon), \\
\hat{p}_0 &\equiv \partial U^\delta \cap W^u_{loc}(p_0),
\end{aligned}
$$

such that

$$|\hat{p}_\epsilon - \hat{p}_0| = \mathcal{O}(\sqrt{\epsilon}).$$

(See Figure 8.5.)

Since $W^u(p_\epsilon) \subset W^s(\mathcal{A}_\epsilon)$, the trajectory through the point $\hat{p}_\epsilon$ at $t = 0$ will return to U^δ some after some *finite* time of flight. We denote this point by

$$p_\epsilon^T \equiv \partial U^\delta \cap W^u(p_\epsilon).$$

From the unperturbed problem we also have a point

$$p_0^T \equiv \partial U^\delta \cap W^u(p_0).$$

Since this time of flight from ∂U^δ to ∂U^δ is finite, it follows from simple Gronwall type estimates that

$$|p_0^T - p_\epsilon^T| = \mathcal{O}(\sqrt{\epsilon}).$$

(See Figure 8.6.)

At this point, we have shown the existence of an orbit that leaves a neighborhood of $\mathcal{A}_\epsilon$ and returns to a neighborhood of $\mathcal{A}_\epsilon$ by using perturbation theory of normally hyperbolic invariant manifolds coupled with a generalized Melnikov type analysis. It remains to show that the orbit approaches p_ϵ asymptotically as $t \to \infty$; to show this, we must use the fibers. It follows from Theorems 8.3.2 and 8.4.1 that the points p_0^T and p_ϵ^T are on fibers. We denote the base points of these fibers by p_0^∞ and p_ϵ^∞, respectively, where, by Theorem 8.4.1

$$|p_0^\infty - p_\epsilon^\infty| = \mathcal{O}(\sqrt{\epsilon}).$$

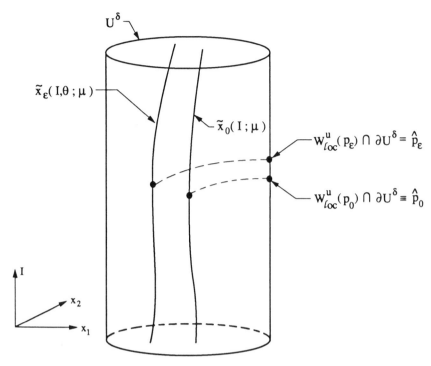

Figure 8.5. The geometry associated with trajectories leaving a neighborhood of $\mathcal{M}_\epsilon$, i.e., the points $\hat{p}_\epsilon$ and $\hat{p}_0$, (with the θ coordinates suppressed).

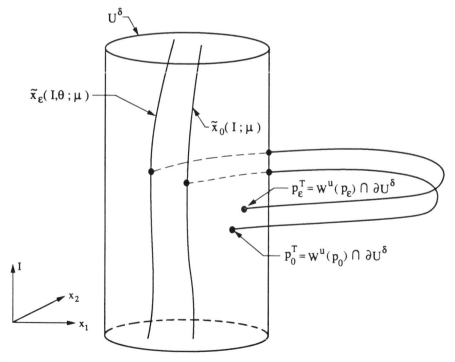

Figure 8.6. The geometry associated with trajectories returning to a neighborhood of $\mathcal{M}_\epsilon$, i.e., the points p_ϵ^T and p_0^T, (with the θ coordinates suppressed).

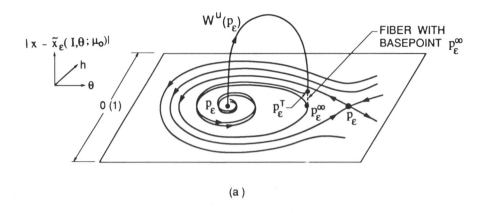

(a)

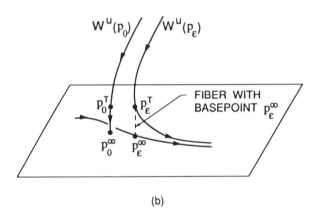

(b)

Figure 8.7. a) The geometry associated with the return of a trajectory to a neighborhood of $\mathcal{A}_\epsilon$ that is homoclinic to p_ϵ. b) The perturbed and unperturbed fibers.

Moreover, the $h - \theta$ coordinates of p_0^∞ are given by

$$p_0^\infty \equiv (0, \theta^\infty(\mu_0)) = (0, \theta_c(\mu_0) + \Delta\theta(\mu_0)),$$

where $\Delta\theta$ is given by (8.8). Now, within $\mathcal{A}_\epsilon$, the domain of attraction of p_ϵ is approximated by the level set of the Hamiltonian connecting q_0 in the sense described in Lemma 8.3.1. Moreover, in the $h - \theta$ coordinates, the area enclosed by this level set of the Hamiltonian is $\mathcal{O}(1)$. Hence, it follows from Lemma 8.3.2 that for ϵ sufficiently small, if p_0^∞ is contained within the homoclinic loop connecting q_0 (and an $\mathcal{O}(1)$ distance away from the homoclinic loop) that p_0^∞ will approach p_ϵ asymptotically as $t \to \infty$. It then follows from Theorem 8.4.1 that the trajectory through p_ϵ^T will approach p_ϵ asymptotically as $t \to \infty$. (See Figure 8.7.)

Verifying this situation is straightforward. Recall from (8.25) that the equation for the homoclinic loop connecting q_0 is given by

$$\mathcal{H}(h, \theta, \mu) - \mathcal{H}(0, \theta_s(\mu_0), \mu_0) = 0.$$

We will refer to the region bounded by this curve as "the fish". In terms of the angle values θ, the "tail" of the fish is at $\theta = \theta_s(\mu_0)$ and the "nose" of the fish is located at $\theta = \theta_n(\mu_0)$ where $\theta_n(\mu_0)$ is found by solving the equation

$$\mathcal{H}(0, \theta, \mu_0) - \mathcal{H}(0, \theta_s(\mu_0), \mu_0) = 0.$$

In Figure 8.8, we illustrate the four general possibilities under our assumptions, which we refer to as cases a to d. We summarize our results in the following theorem.

Theorem 8.4.2. *Suppose that at* $\mu = \mu_0$,

1. $M(I^r, \theta_c(\mu_0), \mu_0) = 0$,

2. $\frac{d}{d\mu} M(I^r, \theta_c(\mu_0), \mu_0)$ *has rank 1,*

with **one** *of the following cases holding (refer to Figure 8.8 for an explanation of the different cases; all angles are taken mod 2π)*

a. $\theta_n(\mu_0) < \theta^\infty(\mu_0) < \theta_s(\mu_0)$,

b. $0 \leq \theta^\infty(\mu_0) < \theta_s(\mu_0)$ **or** $\theta_n(\mu_0) < \theta^\infty(\mu_0) \leq 2\pi$,

c. $\theta_s(\mu_0) < \theta^\infty(\mu_0) < \theta_n(\mu_0)$,

d. $0 \leq \theta^\infty(\mu_0) < \theta_n(\mu_0)$ **or** $\theta_s(\mu_0) < \theta^\infty(\mu_0) \leq 2\pi$,

then (8.1) possesses a "simple" homoclinic orbit connecting p_ϵ. Moreover, if 1 does not hold, then there are no homoclinic orbits connecting p_ϵ and if 1 and the angle inequalities for the appropriate case a-d do not hold, then there are no "simple" homoclinic orbits connecting to p_ϵ.

By a "simple homoclinic orbit" we mean a homoclinic orbit that makes one excursion through a neighborhood of the annulus before connecting p_ϵ. Melnikov type methods say nothing about the existence of homoclinic orbits that may make several passes near the annulus before connecting p_ϵ (we will comment more on this at the end of this chapter).

If all the conditions of this procedure are satisfied, then one has shown the existence of an orbit homoclinic to a fixed point of saddle-focus type (often called a "Silnikov" type homoclinic orbit) in a four-dimensional set of ordinary differential equations. Such orbits are of significance because they give rise to chaotic dynamics in the sense of Smale horseshoes; this is discussed in more detail in Wiggins [1988a] and Kovačič and Wiggins [1992]. Moreover, this method is of importance in engineering applications since it allows for the prediction of specific parameter values for which this chaotic motion arises in weakly damped systems.

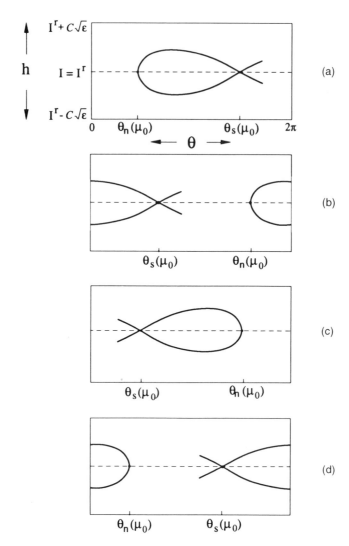

Figure 8.8. Possible geometrical configurations for the unperturbed (on the "slow time" scale) homoclinic orbit on $\mathcal{A}_\epsilon$ that connects p_0.

$$\boxed{W^s(\mathcal{A}_\epsilon) \cap W^u(q_\epsilon)}$$

The transversal intersection of these three-dimensional and two-dimensional sets would be a one-dimensional trajectory. However, the trajectory's forward time fate is unknown and requires further analysis. There is also an additional wrinkle that makes this problem much more difficult than those described above. Namely, in the unperturbed problem q_ϵ has one-dimensional stable and unstable manifolds and a two-dimensional center manifold. Thus, under our assumptions, the two-dimensional $W^u(q_\epsilon)$ is created by the perturbation. This gives rise to issues that are of a singular perturbation type and must be dealt with by using the foliation of the stable and unstable manifolds. In fact, $W^u(q_\epsilon)$ must be "built" from the appropriate fibers. We now show how this can be done following McLaughlin *et al.* [1993].

Fiber Representations for Subsets of $W^u(q_\epsilon)$ and $W^s_{loc}(\mathcal{A}_\epsilon)$

Using regular perturbation arguments, sufficiently close to the fixed point q_ϵ, and for ϵ sufficiently small, the manifold $W^u(q_\epsilon)$, restricted to the annulus $\mathcal{A}_\epsilon \subset \mathcal{M}_\epsilon$, can be represented as a curve in the plane given by h as a function of θ:

$$h^\epsilon_u(\theta) = h_u(\theta) + \mathcal{O}(\sqrt{\epsilon}), \tag{8.58}$$

where h_u is obtained from the equation for the separatrix:

$$\mathcal{H}(h_u(\theta), \theta, \mu_0)) = \mathcal{H}(0, \theta_s(\mu_0), \mu_0). \tag{8.59}$$

(For simplicity of notation, we have left out the explicit dependence of the function $h^\epsilon_u(\theta)$ on the parameter μ.) We explicitly denote the domain of the function $h^\epsilon_u(\theta)$ by

$$(\theta_l < \theta < \theta_r),$$

where $\theta_l \equiv \theta_r - \zeta$, and the constant ζ is chosen such that $h \ll \zeta \ll 1$. The justification for this definition of θ_l is that in the $h - \theta$ coordinates the stable and unstable manifolds of q_ϵ, restricted to $\mathcal{A}_\epsilon$, split by an $\mathcal{O}(\sqrt{\epsilon})$ amount under perturbation. We further define $\theta_r \equiv \theta_s(\mu_0) + C$, where C is any constant satisfying $\theta_s(\mu_0) - C < \theta_n(\mu_0) + \zeta$. (See Figure 8.9.) (Note: if the fish is oriented in the reverse manner then the obvious modifications are made.)

In the $I - \theta$ coordinates, we have

$$\left[W^u(q_\epsilon) \cap \mathcal{M}_\epsilon \right] \supset \{ (\theta, I^\epsilon_u(\theta)), \ \theta \in (\theta_l, \theta_r) \}. \tag{8.60}$$

Here the function $I^\epsilon_u(\theta)$ has the form

$$I^\epsilon_u(\theta) = I^r + \sqrt{\epsilon} h_u(\theta) + \mathcal{O}(\epsilon). \tag{8.61}$$

We can now construct a piece of $W^u(q_\epsilon)$ in the full four-dimensional phase space from the unstable fibers by restricting the base points of the fibers to lie on this curve $I^\epsilon_u(\theta)$ in $\mathcal{M}_\epsilon$. As a general notation, we will denote the unstable fiber with

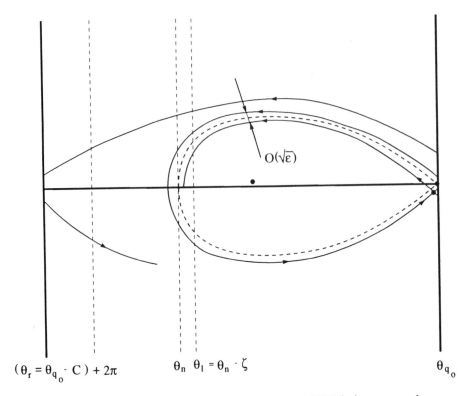

$(\theta_r = \theta_{q_0} \cdot C) + 2\pi$ θ_n $\theta_l = \theta_n \cdot \zeta$ θ_{q_0}

Figure 8.9. Local representation of a segment of $W^u(q_\epsilon)$ as a graph.

base point z_b^ϵ by $\mathcal{F}^{(u,\epsilon)}\{z_b^\epsilon \equiv (\tilde{x}_\epsilon(I_u^\epsilon(\theta),\theta;\mu_0),I_u^\epsilon(\theta),\theta)\}$, similarly for the stable fibers. The resulting fiber representation of this piece of $W^u(q_\epsilon)$ is expressed as follows:

$$W^u(q_\epsilon) \supset \bigcup_{\theta \in (\theta_l, \theta_r)} \mathcal{F}^{(u,\epsilon)}\{z_b^\epsilon \equiv (\tilde{x}_\epsilon(I_u^\epsilon(\theta),\theta;\mu_0),I_u^\epsilon(\theta),\theta)\} \equiv W_{res}^u(q_\epsilon). \quad (8.62)$$

We also define the following sets:

$$\bigcup_{\theta \in (\theta_l, \theta_r)} \mathcal{F}^{(u,\epsilon)}\{z_b^\epsilon \equiv (\tilde{x}_\epsilon(I^r + \sqrt{\epsilon}h_u(\theta),\theta;\mu),I^r + \sqrt{\epsilon}h_u(\theta),\theta)\} \equiv W_{res}^u(q_0),$$

$$(8.63)$$

and

$$\bigcup_{\theta \in (\theta_l, \theta_r)} \mathcal{F}^{(u,0)}\{z_b^0 = (\tilde{x}_0(I^r;\mu_0),I^r,\theta)\} \equiv W_{res}^u(\tilde{x}_0(I^r;\mu_0),I^r,\theta). \quad (8.64)$$

$W_{res}^u(\tilde{x}_0(I^r;\mu),I = I^r,\theta)$ denotes a piece of the two-dimensional unstable manifold of the circle of fixed points in the unperturbed problem. Consider the following cylinder containing $\mathcal{A}_\epsilon$ in the full four-dimensional phase space

$$\mathcal{C}^\delta \equiv \{ z = (x,I,\theta) \mid |x - \tilde{x}_\epsilon(I,\theta,\mu_0)| = \delta^2; \ I,\theta \text{ free} \},$$

where $0 < \delta << 1$ and δ is independent of ϵ. The sets $W_{res}^u(q_\epsilon) \cap \mathcal{C}^\delta$, $W_{res}^u(q_0) \cap \mathcal{C}^\delta$, and $W_{res}^u(\tilde{x}_0(I^r;\mu_0),I = I^r,\theta) \cap \mathcal{C}^\delta$ are curves and are shown in Figure 8.10.

We remark that for δ and ϵ sufficiently small a given fiber intersects $\mathcal{C}^\delta$ transversely in a unique point. This can be computed directly for $\epsilon = 0$ from the expression for the unstable fiber given in McLaughlin *et al.* [1993]. The statement then follows for ϵ sufficiently small by the persistence of transversal intersections. It follows from the perturbation theory for fibers (Theorem 8.3.2) that $W_{res}^u(q_\epsilon) \cap \mathcal{C}^\delta$ and $W_{res}^u(q_0) \cap \mathcal{C}^\delta$ are $C^{(r-1)}$ $\sqrt{\epsilon}$-close to $W_{res}^u(\tilde{x}_\epsilon(I,\theta;\mu),I = I^r,\theta) \cap \mathcal{C}^\delta$. Thus, we see that an important role of the fibers is to "lift" segments of perturbed stable and unstable manifolds off of $\mathcal{A}_\epsilon$ in such a way that we can estimate the closeness of these segments in the full four-dimensional phase space. Clearly, trajectories of solutions could not be used for this purpose because of the singular perturbation nature near the resonance.

Similarly, restricting the base points of the stable fibers to reside in a fixed region $\mathcal{A}_\epsilon$ of the plane $\mathcal{M}_\epsilon$, we can obtain a fiber representation of the local stable manifold of $\mathcal{A}_\epsilon$:

$$W_{loc}^s(\mathcal{A}_\epsilon) \supset \bigcup_{z_b^\epsilon \in \mathcal{A}_\epsilon} \mathcal{F}^{(s,\epsilon)}(z_b^\epsilon). \quad (8.65)$$

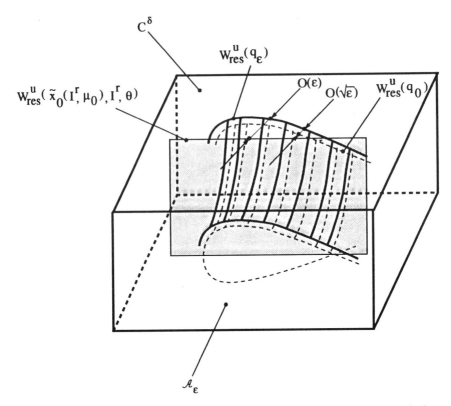

Figure 8.10.   The fiber representation of the sets $W^u_{\text{res}}(q_\epsilon)$, $W^u_{\text{res}}(q_0)$, and $W^u_{\text{res}}\tilde{x}_\epsilon(I,\theta;\mu), I = I^r, \theta)$ in $\mathcal{C}^\delta$.

The Distance Between $W^s(\mathcal{A}_\epsilon)$ and $W^u(q_\epsilon)$

Once fiber representation of the sets $W^u_{\mathrm{res}}(q_\epsilon)$, $W^u_{\mathrm{res}}(q_0)$, and $W^u_{\mathrm{res}}(\tilde{x}_\epsilon(I^r,\theta;\mu))$, $I^r,\theta)$ we can then deal with them in a perturbative fashion. In particular, we want to develop a Melnikov type measurement for the distance between $W^s(\mathcal{A}_\epsilon)$ and the subset of $W^u(q_\epsilon)$ that we constructed with the unstable fibers. This is done in McLaughlin *et al.* [1993] where the appropriate distance measurement is shown to be given by

$$d(I^\epsilon_u(\theta_b),\theta_b,\mu,\epsilon) = \epsilon\frac{M(I^\epsilon_u(\theta_b),\theta_b,\mu)}{\|\,n(p)\,\|} + \mathcal{O}(\epsilon^2),$$

where

$$M(I^\epsilon_u(\theta_b),\theta_b,\mu)$$

$$= \int_{-\infty}^{+\infty} \langle n, (g^x, g^I, g^\theta)\rangle \, (q_0\,(t, I^\epsilon_u(\theta_b),\theta_b,\mu)\,,\mu,0)\,dt,$$

where

$$q_0\,(t, I^\epsilon_u(\theta_b),\theta_b,\mu) \;=\; \left(x^h(t, I^\epsilon_u(\theta_b),\mu), I^\epsilon_u(\theta_b),\right.$$

$$\left. \int_{-\infty}^{t} D_I H(x^h(s, I^\epsilon_u(\theta_b)), I^\epsilon_u(\theta_b),\mu)ds + \theta_b \,\right).$$

The interpretation of the points $(I^\epsilon_u(\theta_b),\theta_b)$ is very important. $I^\epsilon_u(\theta_b)$ is given by (8.61). In other words, $(I^\epsilon_u(\theta_b),\theta_b)$ is constrained to lie on a segment of $W^u(q_\epsilon) \subset \mathcal{A}_\epsilon$ (the "back of the fish"). The subscript b has been added to emphasize that these points are to be regarded as base points of fibers in the foliation of the unstable manifold. Expanding the distance, as well as its argument (8.61), in powers of $\sqrt{\epsilon}$, the leading order term (neglecting the nonzero normalization factor $\|\,n(p)\,\|$) becomes

$$M(I^r,\theta_b;\mu)$$

$$= \int_{-\infty}^{+\infty} \langle n, (g^x, g^I, g^\theta)\rangle \left(x^h(t, I^r,\mu), I^r, \int_{-\infty}^{t} D_I H(x^h(s, I^r)), I^r,\mu)ds + \theta_b,\mu,0 \right) dt.$$

$$(8.66)$$

Fortunately, for verifying the existence of orbits homoclinic to q_ϵ, this latter expression will suffice.

Recall that the transversal intersection of the three-dimensional $W^s(\mathcal{A}_\epsilon)$ and the two-dimensional $W^u(q_\epsilon)$ is a one-dimensional trajectory. It is also important to recall the interpretation of $I^\epsilon_u(\theta_b)$ and θ_b given in the previous section; namely, they label the base point of a fiber in $W^u(q_\epsilon)$ having the property that the trajectory in $W^s(\mathcal{A}_\epsilon) \cap W^u(q_\epsilon)$ corresponding to the zero of the Melnikov function asymptotes *in negative time* to a trajectory in $\mathcal{A}_\epsilon \cap W^u(q_\epsilon)$ that passes through the point $(I^\epsilon_u(\theta_b),\theta_b)$. Our goal now is to show that under certain conditions this trajectory asymptotes to a trajectory in $\mathcal{A}_\epsilon \cap W^s(q_\epsilon)$. In other words, q_ϵ has a homoclinic orbit. We begin by discussing the base points of the fibers of $W^s(\mathcal{A}_\epsilon)$ in terms of the geometry of trajectories in $\mathcal{A}_\epsilon$.

Base Points of Fibers and the Dynamics on $\mathcal{A}_\epsilon$

We want to describe the location of base points of fibers of $W^s(\mathcal{A}_\epsilon)$ in terms of their location with respect to the level sets of the Hamiltonian (8.24). At a zero of the Melnikov function, let the corresponding initial condition $W^s(\mathcal{A}_\epsilon)$ flow forward in time along the perturbed trajectory until it reaches $\mathcal{C}^\delta$. Then the point can be considered to be on a stable fiber, $\mathcal{F}^{(s,\epsilon)}(z_b^\epsilon) \subset W^s_{loc}(\mathcal{A}_\epsilon)$. Let $\left(\tilde{h}_b^L, \tilde{\theta}_b^L\right)$ denote the base point of this fiber. (Mnemonically, the superscript "L" stands for "landing point".) Also recall that, by construction, $(I_u(\theta_b), \theta_b)$ lies on $W^u(q_\epsilon)$. Then the following quantity,

$$\Delta\mathcal{H} \equiv \mathcal{H}(\tilde{h}_b^L, \tilde{\theta}_b^L, \mu) - \mathcal{H}(I_u(\theta_b), \theta_b, \mu), \tag{8.67}$$

indicates whether or not $\left(\tilde{h}_b^L, \tilde{\theta}_b^L\right)$ lies inside or outside this homoclinic orbit defined by $\mathcal{H}(h, \theta, \mu) - \mathcal{H}(0, \theta_s(\mu), \mu)$. Suppose the values of the level sets of (8.24) are monotone increasing as one moves outwards from p_ϵ (other cases are equally easy to consider), then $\Delta\mathcal{H} > 0$ implies that $\left(\tilde{h}_b^L, \tilde{\theta}_b^L\right)$ lies outside the homoclinic orbit and $\Delta\mathcal{H} < 0$ implies that $\left(\tilde{h}_b^L, \tilde{\theta}_b^L\right)$ lies inside the homoclinic orbit. (See Figures 8.11a,b.)

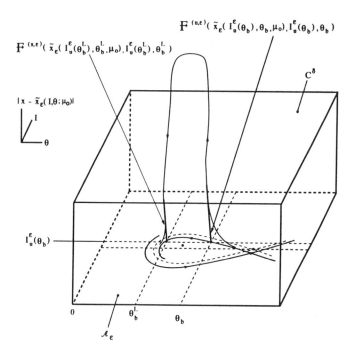

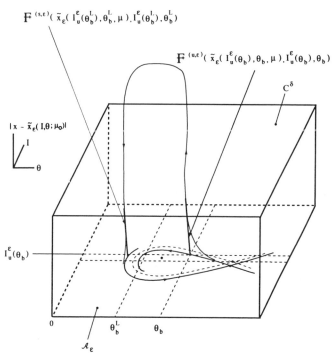

Figure 8.11. a) $\Delta\mathcal{H} < 0$ b) $\Delta\mathcal{H} > 0$ c) $\Delta\mathcal{H} \approx 0$ (*Continued on next page.*)

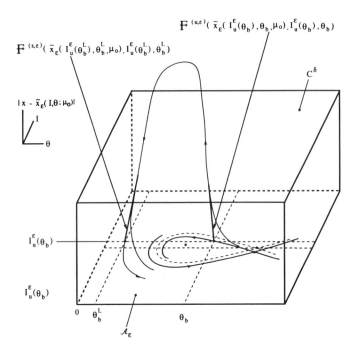

The following Lemma allows us to explicitly compute this criteria.

Lemma 8.4.1.

$$\left(\tilde{h}_b^L, \tilde{\theta}_b^L\right) = \left(h_u(\theta_b) \equiv h_b^L, \theta_b^L\right) + \mathcal{O}(\sqrt{\epsilon}),$$

where $h_u(\theta_b^L)$ is given by the solution of (8.59), $\theta_b^L = \theta_b + \Delta\theta(\mu_0)$, and $\Delta\theta(\mu_0)$ is given by (8.8).

Proof: From arguments given above, the perturbed trajectory enters $\mathcal{C}^\delta$ $\mathcal{O}\sqrt{\epsilon}$-close to an unperturbed orbit that connects points on the circle of fixed points at $I = I^r$. This result then follows from Fenichel's perturbation theory for stable fibers.   □

The Shooting Argument: The Existence of an Orbit Homoclinic to q_ϵ

Now it is important to understand the use of the expression $\Delta\mathcal{H}$. A simple zero of the Melnikov function corresponds to a trajectory in $W^s(\mathcal{A}_\epsilon) \cap W^u(q_\epsilon)$; however, the $\Delta\mathcal{H}$ value at this simple zero gives us information concerning the positive time asymptotic behavior of this trajectory. We describe this situation in more detail. We are interested in situations when the homoclinic trajectory asymptotes to orbits that are inside and outside the domain of attraction of p_ϵ. In terms of the dynamics on $\mathcal{M}_\epsilon$, this depends upon whether or not (h_b^L, θ_b^L) is inside the domain of attraction of p_ϵ, which is $\mathcal{A}_\epsilon \cap W^s(q_\epsilon)$. Moreover, because of the $\mathcal{O}(\sqrt{\epsilon})$ error in the actual location of the base point of the fiber, we will also impose the sufficient condition that (h_b^L, θ_b^L) is an $\mathcal{O}(1)$ distance from the boundary of the basin of attraction of p_ϵ. By the interpretation given above, for a zero of the Melnikov function with $\mathcal{H} > 0$ and $\mathcal{O}(1)$, the corresponding trajectory in $W^s(\mathcal{A}_\epsilon) \cap W^u(q_\epsilon)$ is asymptotic to a trajectory on $\mathcal{A}_\epsilon$ that is not in the domain of attraction of p_ϵ, and for $\mathcal{H} < 0$ and $\mathcal{O}(1)$ the corresponding trajectory in $W^s(\mathcal{A}_\epsilon) \cap W^u(q_\epsilon)$ is asymptotic to a trajectory on $\mathcal{A}_\epsilon$ that is in the domain of attraction of p_ϵ. Hence, since all manifolds, as well as the Melnikov function, depend continuously on the parameters, there must be some intermediate value of $\mathcal{H}$ such that the trajectory corresponding to the zero of the Melnikov function asymptotes to a trajectory in $W^s(q_\epsilon) \cap \mathcal{A}_\epsilon$, i.e., the boundary of the domain of attraction of p_ϵ. Thus we have the existence of an orbit homoclinic to q_ϵ. (See Figure 8.11.)

In order for this to hold it is sufficient for $\mathcal{H}$ to take values throughout an $\mathcal{O}(1)$ interval about $\mathcal{H} = 0$, where also on this interval the Melnikov function has a simple zero. We summarize our results in the following theorem from McLaughlin *et al.* [1993].

Theorem 8.4.3. *Suppose that on the set of points for which*

$$M(I^r, \theta_b; \mu_0) = 0$$

has simple zeros, the quantity

$$\Delta\mathcal{H} = \mathcal{H}(h_u(\theta_b), \theta_b + \Delta\theta(\mu_0), \mu_0) - \mathcal{H}(h_u(\theta_b), \theta_b, \mu_0)$$

can take values in an $\mathcal{O}(1)$ size neighborhood about 0. Then $W^s(q_\epsilon)$ intersects $W^u(q_\epsilon)$.

Concerning chaotic dynamics, homoclinic orbits associated with saddle points having purely real eigenvalues need not give rise to chaotic dynamics of the Smale horseshoe type. However, if there are a pair of homoclinic orbits (such as in the case where the vector field is symmetric), then it may be possible to carry out a horseshoe construction in the neighborhood of the homoclinic orbits. Of course, the most famous example of this is the Lorenz equations in three dimensions. In four dimensions, this phenomenon is considered in McLaughlin *et al.* [1993]. Finally, we remark that $W^s(q_\epsilon) \cap W^u(q_\epsilon)$ is a codimension two phenomenon whereas $W^s(p_\epsilon) \cap W^u(p_\epsilon)$ is a codimension one phenomenon.

Orbits Homoclinic and Heteroclinic to Internal Orbits: Hamiltonian Perturbations

The difficulties with this case arise due to the fact that in the unperturbed problem, all of the resonant orbits have the same energy and their stable and unstable manifolds intersect non-transversely. The analysis involves a number of delicate considerations involving energy conservation and transversality that crucially uses the fibers in the foliation of $W^s(\mathcal{A}_\epsilon)$ and $W^u(\mathcal{A}_\epsilon)$. This next section summarizes the key results of Haller and Wiggins [1993a].

The Intersection of $\mathcal{A}_\epsilon$, $W^s(\mathcal{A}_\epsilon)$ and $W^u(\mathcal{A}_\epsilon)$ with $E_\epsilon(h)$

First we define

$$E_\epsilon(h) = \{(x, I, \theta) | H(x, I, \theta; \mu, \epsilon) = h\}$$

to be the *perturbed energy surface* with energy h (where $H(x, I, \theta; \mu, \epsilon) = H(x, I, \mu) + \epsilon H_1(x, I, \theta; \mu, \epsilon)$). In the following, we will examine the intersection of $\mathcal{A}_\epsilon$ with cer- tain perturbed energy surfaces. *From this point, D (without subscript) will refer to the gradient operator in the variables (x, I, θ).*
The nature of the intersection of $E_\epsilon(h)$ with internal orbits in $\mathcal{A}_\epsilon$ is described in the following proposition.

Proposition 8.4.1. *Let us assume that γ_0 is an internal orbit of the reduced sys- tem (8.39) with $\mathcal{H}|\gamma_0 = h_0$. Then, for any $p \in \gamma_0$ there exists $\epsilon_0 > 0$ and an open neighborhood $U_p \subset A$ with $p \in U_p$ such that for $0 < \epsilon < \epsilon_0$ the following are satisfied:*

1. *Any orbit $\gamma_\epsilon \in A$ of the restricted system (8.38) with $\gamma_\epsilon \cap U_p \neq \emptyset$ is an internal orbit. Moreover, if $\mathcal{H}_\epsilon|\gamma_\epsilon = h$ then $E_\epsilon(h)$ intersects $\mathcal{A}_\epsilon$ transversally in γ_ϵ.*

2. *There exists an orbit $\gamma_\epsilon^* \in A$ with $\gamma_\epsilon^* \cap U_p \neq \emptyset$ which is C^{r-1} $\sqrt{\epsilon}$-close to γ_0.*

Proof: See Haller and Wiggins [1993a].   □

Using Proposition 8.4.1 we can obtain information about the intersection of certain energy surfaces with $W^s(\mathcal{A}_\epsilon)$ and $W^u(\mathcal{A}_\epsilon)$. We can actually prove the following proposition.

Proposition 8.4.2. *Let us assume that γ_0 is an internal orbit of the reduced sys- tem (8.39). Then, for any $p \in \gamma_0$ there exists $\epsilon_0 > 0$ and an open neighborhood $U_p \subset A$ with $p \in U_p$ such that for $0 < \epsilon < \epsilon_0$ the following are satisfied:*

1. *For any γ_ϵ orbit of the restricted system (8.38) with $\gamma_\epsilon \cap U_p \neq \emptyset$ and $H_\epsilon|\gamma_\epsilon = h$, the energy surface $E_\epsilon(h)$ intersects $W^s(\mathcal{A}_\epsilon)$ and $W^u(\mathcal{A}_\epsilon)$.*

2. *At least one connected component of the intersection $W^s(\mathcal{A}_\epsilon) \cap E_\epsilon(h)$ (respec- tively $W^u(\mathcal{A}_\epsilon) \cap E_\epsilon(h)$) is transversal and identical to $W^s(\gamma_\epsilon)$ (respectively $W^u(\gamma_\epsilon)$).*

3. *For sufficiently small δ_0 with $0 < \epsilon << \delta_0 << 1$, $W^s(\gamma_\epsilon) \cup W^u(\gamma_\epsilon) \cap U^{\delta_0}$ is approximated with an error of $\mathcal{O}(\epsilon, \delta_0^2)$ by the manifold $\tilde{F}_0 \subset \Gamma$ satisfying the following equations:*

$$\begin{aligned} H(x, I^r; \mu) &= H(\tilde{x}_0(I^r; \mu), I^r; \mu), \\ \mathcal{H}_\epsilon(I, \theta; \mu) &= h. \end{aligned}$$

Proof: See Haller and Wiggins [1993a].   □

We make the following remarks concerning these results.

1. In Proposition 8.4.2 we have only asserted that $W^s(\gamma_\epsilon) \subset E_\epsilon(h) \cap W^s(\mathcal{A}_\epsilon)$ and $W^u(\gamma_\epsilon) \subset E_\epsilon(h) \cap W^u(\mathcal{A}_\epsilon)$. In particular, we did not claim that, for example, $W^s(\gamma_\epsilon) = E_\epsilon(h) \cap W^s(\mathcal{A}_\epsilon)$. In fact, $E_\epsilon(h) \cap W^s(\mathcal{A}_\epsilon)$ can have several connected components and precisely one of these connected components is equal to $W^s(\gamma_\epsilon)$.

2. If $\gamma_\epsilon \in \mathcal{A}_\epsilon$ is an orbit homoclinic to some equilibrium $q_\epsilon \in \mathcal{A}_\epsilon$ then by $W^s(\gamma_\epsilon)$ (respectively $W^u(\gamma_\epsilon)$) we mean the connected component of $W^s(q_\epsilon)$ (respectively $W^s(q_\epsilon)$) which contains γ_ϵ.

The Intersection of $W^s(\mathcal{A}_\epsilon)$ and $W^u(\mathcal{A}_\epsilon)$

Let us first define the *energy-difference function* on the annulus A

$$
\begin{aligned}
\Delta\mathcal{H}(\theta; \mu) &= \mathcal{H}(h, \theta; \mu) - \mathcal{H}(h, \theta - \Delta\theta(\mu); \mu), \\
&= H_p(\tilde{x}_0(I^r; \mu), I^r, \theta; \mu) - H_p(\tilde{x}_0(I^r; \mu), I^r, \theta - \Delta\theta(\mu); \mu), \quad (8.68)
\end{aligned}
$$

with $\Delta\theta(\mu)$ defined in (8.8) and H_p defined in (8.34). The set

$$
Z_\mu^+ = \{(h, \theta) \in A | \Delta\mathcal{H}(\theta; \mu) = 0, \quad D_\theta \Delta\mathcal{H}(\theta; \mu) \neq 0\}, \qquad (8.69)
$$

which is the set of transversal zeros of $\Delta\mathcal{H}$. Note that if Z_μ^+ is not empty, then it consists of a set of $\theta = const.$ lines.

We will now formulate a result concerning the intersection of $W^s(\mathcal{A}_\epsilon)$ and $W^u(\mathcal{A}_\epsilon)$ in terms of $\Delta\mathcal{H}$ and $\Delta\theta$.

Lemma 8.4.2. *Let us suppose that there exists $\mu_0 \in V$ such that $Z_{\mu_0}^+$ is not empty and $p = (\bar{h}, \bar{\theta}) \in Z_{\mu_0}^+$. Then, there exists ϵ_0 such that if $\epsilon < \epsilon_0$ then*

1. *$W^s(\mathcal{A}_\epsilon) \cap W^u(\mathcal{A}_\epsilon) = \Gamma_\epsilon \neq \emptyset$.*

2. *The intersection of $W^s(\mathcal{A}_\epsilon)$ and $W^u(\mathcal{A}_\epsilon)$ is transversal along an orbit.*

Proof: The proof is taken from Haller and Wiggins [1993a]. We first recall that Γ satisfies the equation

$$
H(x, I^r; \mu) = H(\tilde{x}_0(I^r; \mu), I^r; \mu).
$$

This, together with the parametrization of Γ in terms of the unperturbed trajectories shows that a normal to Γ at a point described by the parameters (t_0, I, θ_0) is given by

$$
n(-t_0, I, \theta_0; \mu) = DH(x^h(-t_0, I; \mu), I; \mu),
$$

where D again refers to the gradient taken in the variables (x, I, θ). For small $\epsilon > 0$, both $W^s(\mathcal{A}_\epsilon)$ and $W^u(\mathcal{A}_\epsilon)$ intersect n transversally in two points, q_ϵ^s and q_ϵ^u, respectively, which are locally unique and may coincide. We define the signed splitting distance of $W^s(\mathcal{A}_\epsilon)$ and $W^u(\mathcal{A}_\epsilon)$ at (t_0, I, θ_0) as

$$d(t_0, I, \theta_0; \mu, \epsilon) = \frac{\langle n(-t_0, I, \theta_0; \mu), q_\epsilon^s - q_\epsilon^u \rangle}{\|n(-t_0, I, \theta_0; \mu)\|}.$$

It can be shown (see the references mentioned above) that d can be written as

$$d(t_0, I, \theta_0; \mu, \epsilon) = \epsilon \frac{M(t_0, I, \theta_0; \mu)}{\|n(-t_0, I, \theta_0; \mu)\|} + \mathcal{O}(\epsilon^2), \tag{8.70}$$

where M is the *Melnikov function* given by

$$M(t_0, I, \theta_0; \mu) = \int_{-\infty}^{+\infty} \{H, H_p\}|_{q_0(t - t_0, I, \theta_0; \mu)} \, dt = \int_{-\infty}^{+\infty} \frac{d}{dt} H_p|_{q_0(t - t_0, I, \theta_0; \mu)} \, dt, \tag{8.71}$$

with the solution q_0 in the form as in (8.53)and $\{\,,\,\}$ denoting the canonical Poisson-bracket. Evaluating (5.11) we find that

$$M(t_0, I, \theta_0; \mu) = H_p|_{q_0(+\infty, I, \theta_0; \mu)} - H_p|_{q_0(-\infty, I, \theta_0; \mu)}$$

$$= H_p(\tilde{x}_0(I^r, \mu), I^r + \sqrt{\epsilon}\bar{h}, \bar{\theta}; \mu) - H_p(\tilde{x}_0(I^r, \mu), I^r + \sqrt{\epsilon}\bar{h}, \bar{\theta} - \Delta\theta(\mu); \mu), \tag{8.72}$$

where we have introduced the notation

$$\bar{\theta} = \theta_0 + \int_{t_0}^{+\infty} D_I H(x^h(t, I^r; \mu), I^r; \mu) \, dt. \tag{8.73}$$

For small ϵ, we Taylor-expand (8.72) to obtain

$$M(t_0, I, \theta_0; \mu) = \Delta\mathcal{H}(\bar{\theta}; \mu) + \mathcal{O}(\sqrt{\epsilon}). \tag{8.74}$$

Using (8.70) and the implicit function theorem, we see that if $D_\theta \Delta\mathcal{H}(\bar{\theta}; \mu) \neq 0$ then for $\mu = \mu_0$, small positive ϵ and any $I = \bar{I} = I^r + \sqrt{\epsilon}\bar{h} \in [\bar{I}_1, \bar{I}_2]$, the distance function in (8.70) has a transverse zero $\mathcal{O}(\epsilon)$-close to $(t_0(\bar{\theta}), \bar{I}, \theta_0(\bar{\theta}))$. This means that $W^s(\mathcal{A}_\epsilon)$ and $W^u(\mathcal{A}_\epsilon)$ intersect transversally in an orbit y_ϵ which has a point $\mathcal{O}(\epsilon)$-close to the point of Γ_0 with parameters $(\bar{t}_0, \bar{I}, \bar{\theta}_0)$. Based on this, (8.68) and (8.72) to (8.74) the lemma is proved.   □

Orbits Asymptotic to Internal Orbits: The Energy-Phase Criterion

We are now at the point where we can state the main result. It is important to note that we will only use three ingredients in formulating this result:

- Our knowledge about the unperturbed geometry of the standard form, embodied in the phase shift $\Delta\theta$.

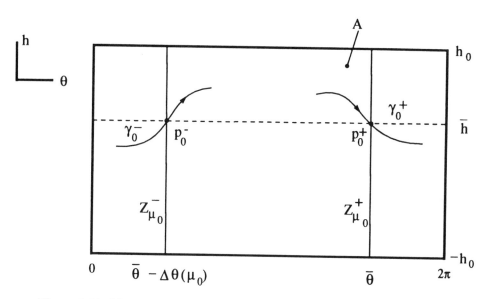

Figure 8.12. Visualization of assumptions $(A1)\,to\,(A3)$ in Theorem 8.4.4.

- The dynamics of the reduced system.

- The structure of the transverse zero set Z_μ^+ of the energy difference function $\Delta\mathcal{H}$, and the structure of the set

$$Z_\mu^- = \{(h,\theta) \in A| \ (h,\theta + \Delta\theta(\mu)) \in Z_\mu^+\}, \qquad (8.75)$$

which is just the counterpart of Z_μ^+ in the following sense: trajectories, which are asymptotic to a point of Z_μ^+ in forward time, are asymptotic to the corresponding point in Z_μ^- in backward time. We remark that $Z_\mu^+ \neq \emptyset$ implies $Z_\mu^- \neq \emptyset$.

Finally, we emphasize that *all the distances mentioned in Theorem 8.4.4 are measured in the coordinates* (x, I, θ).

Theorem 8.4.4. *Let us assume that assumption 1 and assumption 2 hold, and (see Figure 8.12)*

(A1) There exists $\mu_0 \in V$ such that $Z_{\mu_0}^+ \neq \emptyset$,

(A2) γ_0^+ and γ_0^- are internal orbits of the reduced system (8.39) with $\gamma_0^+ \cap Z_{\mu_0}^+ = p_0^+ = (\bar{h}, \bar{\theta}) \in A$ and $\gamma_0^- \cap Z_{\mu_0}^- = p_0^- = (\bar{h}, \bar{\theta} - \Delta\theta(\mu_0)) \in A$,

(A3) γ_0^+ intersects $Z_{\mu_0}^+$ transversally at p_0^+ and and γ_0^- intersect $Z_{\mu_0}^-$ transversally at p_0^-.

Then, there exists $\epsilon_0 > 0$ such that for $0 < \epsilon < \epsilon_0$ the following are satisfied:

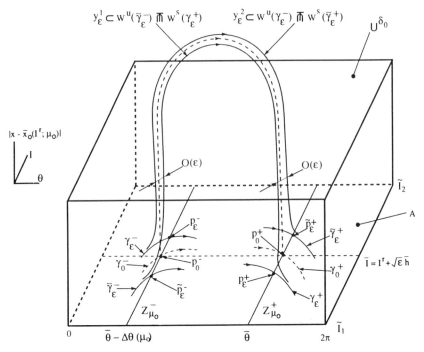

Figure 8.13. Statement 2 of Theorem 8.4.4.

1. The perturbed system has two internal orbits, $\gamma_\epsilon^+, \gamma_\epsilon^- \in \mathcal{A}_\epsilon$ which are C^{r-1} $\sqrt{\epsilon}$-close in A to γ_0^+ and γ_0^-, respectively.

2. The perturbed system also has two internal orbits, $\tilde{\gamma}_\epsilon^+, \tilde{\gamma}_\epsilon^- \in \mathcal{A}_\epsilon$ such that $\tilde{\gamma}_\epsilon^+$ intersects $Z_{\mu_0}^+$ transversally at $\tilde{p}_\epsilon^+$ with $d(\tilde{p}_\epsilon^+, p_0^+) = \mathcal{O}(\sqrt{\epsilon})$ in A, and $\tilde{\gamma}_\epsilon^-$ intersects $Z_{\mu_0}^-$ transversally at $\tilde{p}_\epsilon^-$ with $d(\tilde{p}_\epsilon^-, p_0^-) = \mathcal{O}(\sqrt{\epsilon})$ in A. Moreover, $W^s(\gamma_\epsilon^+)$ intersects $W^u(\tilde{\gamma}_\epsilon^-)$ and $W^u(\gamma_\epsilon^-)$ intersects $W^s(\tilde{\gamma}_\epsilon^+)$ in transverse heteroclinic orbits y_ϵ^1 and y_ϵ^2, respectively. (See Figure 8.13.)

3. If γ_0^+ and γ_0^- are distinct periodic orbits, then y_ϵ^1 and y_ϵ^2 are transverse heteroclinic orbits connecting periodic solutions. (See Figure 8.14.) If $\gamma_0^+ = \gamma_0^-$ is a periodic orbit then $y_\epsilon^1 = y_\epsilon^2$ is a transverse homoclinic orbit connecting a periodic solution to itself.

4. If γ_0^+ and γ_0^- are distinct homoclinic orbits then y_ϵ^1 and y_ϵ^1 are generically transverse heteroclinic orbits connecting saddle points to periodic solutions, but in degenerate (or symmetric) cases they might be transverse heteroclinic orbits connecting saddle points. If $\gamma_0^+ = \gamma_0^-$ is a homoclinic orbit, then $y_\epsilon^1 = y_\epsilon^2$ is a transverse homoclinic orbit connecting a saddle point to itself. (See Figure 8.15.)

5. If one of γ_0^+ and γ_0^- is a homoclinic orbit and the other is a periodic orbit, then one of y_ϵ^1 and y_ϵ^2 is a transverse heteroclinic orbit connecting a saddle point to a periodic solution, while the other is a transverse heteroclinic orbit connecting periodic solutions. (See Figure 8.16.)

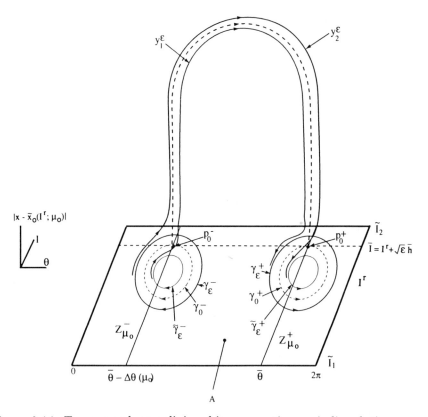

Figure 8.14. Transverse heteroclinic orbits connecting periodic solutions.

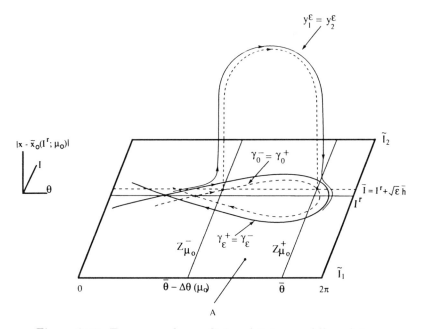

Figure 8.15. Transverse homoclinic orbit to a saddle point.

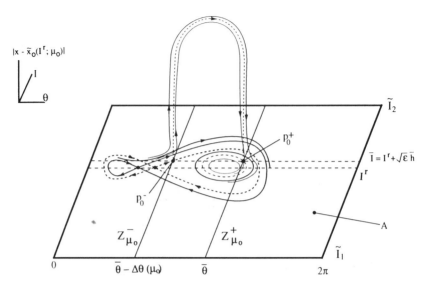

Figure 8.16. Transverse heteroclinic orbits between periodic solutions and between a saddle point and a periodic solution.

Proof: See Haller and Wiggins [1993a].   □

   We note that orbits homoclinic to hyperbolic periodic orbits give rise to Smale horseshoes and their attendant chaotic dynamics. A very detailed discussion of chaos associated with orbits homoclinic to resonances in Hamiltonian systems can be found in Haller and Wiggins [1993a].

8.5. An Example: Dynamics of a Modal Truncation of the Nonlinear Schrödinger Equation

   In this section we consider an example that illustrates all of the previously developed theory. The development of all of the theory in this chapter was motivated by the desire to understand a variety of numerical experiments on the damped and driven Sine-Gordon equation performed by Bishop *et al.* [1986], [1988]. We give a brief description of their results. Consider the perturbed Sine-Gordon equation

$$u_{tt} - u_{xx} + \sin u = \epsilon[-\widehat{\alpha}u_t + \widehat{\Lambda}u_{txx} + \widehat{\Gamma}\sin\omega t],$$

with periodic, even boundary conditions:

$$u\left(x = -\frac{L}{2}, t\right) = u\left(x = \frac{L}{2}, t\right)$$
$$u(x, t) = u(-x, t)$$

with $0 < \epsilon\widehat{\alpha} << 1$, $0 < \epsilon|\widehat{\Lambda}| << 1$, $\omega = 1 - \epsilon\widetilde{\omega}$, and L fixed where ϵ and $\widetilde{\omega}$ are both positive. In the original numerical experiments $\widehat{\Lambda}$ was taken to be zero; however, we will include it in our analysis and discuss the reasons for this when we describe the *two-mode model* at the end of the introduction. In the numerical experiments, Bishop *et al.* chose a single-hump Sine-Gordon "breather" as an initial condition and observed its evolution in space and time as they increased the forcing $\epsilon\widehat{\Gamma}$.

The complete bifurcation sequence is described in Bishop *et al.* [1990a, b]. Here we only present the main point, which is that above a certain value of $\epsilon\widehat{\Gamma}$ (with $\widehat{\Lambda} = 0$), chaotic jumping of the solution occurs between two "breathers", one peaked at the middle and the other at the ends of the interval $[\frac{-L}{2}, \frac{L}{2}]$, with the solution passing near a spatially independent (or "flat") solution on every jump. Comparing this situation with the geometrical structure in the phase space of the *unperturbed* Sine-Gordon equation (Ercolani *et al.* [1990]), they found that the latter possesses linearly unstable spatially independent solutions connected to themselves by homoclinic orbits: the two types of "breathers" with a spatial hump structure which exhibit chaotic behavior in the perturbed problem. From the knowledge of the geometrical structure of the phase space of the unperturbed system, as well as the fact that homoclinic orbits appeared to be involved in the numerically observed chaos, Bishop *et al.* inferred that a Melnikov-type analysis could possibly be developed in order to see how these unperturbed structures are distorted under perturbation in such a way as to give rise to chaos. As a preliminary step in the analysis they chose to develop a simple model that captures the essential structure of the unperturbed and perturbed Sine-Gordon equation. It is this simple model that we now describe.

At small amplitudes, and for frequencies close to 1, the dynamics of the Sine-Gordon equation can be approximated by the nonlinear Schrödinger envelope equation. This can be seen as follows: if we seek a solution of the perturbed Sine-Gordon equation of the form

$$u_\epsilon(x,t) = 2\sqrt{\epsilon\widetilde{\omega}}[B(X,T)e^{i\omega t} + B^*(X,T)e^{-i\omega t}] + \mathcal{O}(\epsilon),$$

with $X = \sqrt{2\epsilon\widetilde{\omega}}x$ and $T = \epsilon\widetilde{\omega}t$, and substitute it into the perturbed Sine-Gordon equation, we obtain the following perturbed nonlinear Schrödinger equation for $B(X,T)$

$$-iB_T + B_{XX} + (|B|^2 - 1)B = \epsilon\left[i\alpha B - i\Lambda B_{XX} + i\bar{\Gamma}\right],$$

where $\widehat{\alpha} = 2\epsilon\widetilde{\omega}\alpha$, $\widehat{\Lambda} = \Lambda$, and $\widehat{\Gamma} = 8\epsilon^{\frac{3}{2}}\widetilde{\omega}^{\frac{3}{2}}\bar{\Gamma}$. This equation is defined on the X interval $[-\frac{L_X}{2}, \frac{L_X}{2}]$ where $L_X = \sqrt{2\epsilon\widetilde{\omega}}L$.

Bishop *et al.* next constructed a two (Fourier) mode truncation to the nonlinear Schrödinger equation by assuming a solution of the form

$$B(X,T) = \frac{1}{\sqrt{2}}c(T) + b(T)\cos kX$$

(with $k = \frac{2\pi}{L_X}$). Substituting this solution into the equation and neglecting the higher Fourier modes gives the following *two-mode model*

$$-i\dot{c} + (\frac{1}{2}|c|^2 + \frac{1}{2}|b|^2 - 1)c + \frac{1}{2}(cb^* + bc^*)b = i\epsilon\alpha c + i\epsilon\Gamma,$$

$$-i\dot{b} + (\frac{1}{2}|c|^2 + \frac{3}{4}|b|^2 - (1 + k^2))b + \frac{1}{2}(cb^* + bc^*)c = i\epsilon\beta b,$$

where we have set $\Gamma = \sqrt{2}\bar{\Gamma}$, and $\beta = (\alpha + \Lambda k^2)$.

If we make the coordinate transformation

$$c = |c|e^{i\theta}, \qquad b = (x + iy)e^{i\theta},$$

then the *perturbed equations* are given by

$$
\begin{aligned}
\dot{x} &= -k^2 y - \frac{3}{4}x^2 y + \frac{1}{4}y^3 + \epsilon\left[\Gamma\frac{y}{\sqrt{2I - x^2 - y^2}}\sin\theta - \beta x\right], \\
\dot{y} &= (k^2 - 2I)x + \frac{7}{4}x^3 + \frac{3}{4}xy^2 - \epsilon\left[\Gamma\frac{x}{\sqrt{2I - x^2 - y^2}}\sin\theta + \beta y\right], \\
\dot{I} &= -\epsilon\left[\Gamma\sqrt{2I - x^2 - y^2}\cos\theta + (\beta - \alpha)(x^2 + y^2) + 2\alpha I\right], \\
\dot{\theta} &= 1 - I - x^2 + \epsilon\Gamma\frac{1}{\sqrt{2I - x^2 - y^2}}\sin\theta,
\end{aligned}
\tag{8.76}
$$

which have the general form for our theory:

$$
\begin{aligned}
\dot{x} &= \frac{\partial H}{\partial y} + \epsilon\frac{\partial H_1}{\partial y} - \epsilon\beta x, \\
\dot{y} &= -\frac{\partial H}{\partial x} - \epsilon\frac{\partial H_1}{\partial x} - \epsilon\beta y, \\
\dot{I} &= \epsilon\frac{\partial H_1}{\partial \theta} - \epsilon 2\alpha I - \epsilon(\beta - \alpha)(x^2 + y^2), \\
\dot{\theta} &= -\frac{\partial H}{\partial I} - \epsilon\frac{\partial H_1}{\partial I},
\end{aligned}
\tag{8.77}
$$

where

$$H = \frac{1}{2}I^2 - I - \frac{7}{16}x^4 - \frac{3}{8}x^2 y^2 + \frac{1}{16}y^4 + (I - \frac{1}{2}k^2)x^2 - \frac{1}{2}k^2 y^2 \tag{8.78}$$

and

$$H_1 = -\Gamma\sqrt{2I - x^2 - y^2}\sin\theta. \tag{8.79}$$

Note that the perturbed equations do not change under the coordinate transformation

$$(x, y, I, \theta) \mapsto (-x, -y, I, \theta).$$

This implies that any homoclinic orbit that we find has a *symmetric partner*.

The *unperturbed equations* are obtained by setting $\epsilon = 0$:

$$
\begin{aligned}
\dot{x} &= -k^2 y - \frac{3}{4}x^2 y + \frac{1}{4}y^3 = \frac{\partial H}{\partial y}, \\
\dot{y} &= (k^2 - 2I)x + \frac{7}{4}x^3 + \frac{3}{4}xy^2 = -\frac{\partial H}{\partial x},
\end{aligned}
$$

$$\dot{I} = 0 = \frac{\partial H}{\partial \theta},$$

$$\dot{\theta} = 1 - I - x^2 = -\frac{\partial H}{\partial I}, \tag{8.80}$$

and are clearly integrable with the two integrals given by H and I. We now proceed to analyze this system using the methods developed in this chapter.

Note that the scalar variables $x - y$ play the role of our general variable $x \in \mathbb{R}^2$ used in the development of the general theory.

8.5.1. The Unperturbed Structure.

$$\boxed{\mathcal{M} \ and \ W^s(\mathcal{M}) \cap W^u(\mathcal{M})}$$

We consider the $x - y$ component of (8.80) which we rewrite below:

$$\dot{x} = -k^2 y - \frac{3}{4}x^2 y + \frac{1}{4}y^3,$$

$$\dot{y} = (k^2 - 2I)x + \frac{7}{4}x^3 + \frac{3}{4}xy^2. \tag{8.81}$$

Note that (8.81) has a fixed point at $(x, y) = (0, 0)$ for all values of I. A simple linear stability analysis shows that $(x, y) = (0, 0)$ is a saddle point for $I > \frac{k^2}{2}$. Moreover, an examination of the level set that contains the origin, i.e., $\{ (x, y) \mid H(x, y, I) = H(0, 0, I) \}$, shows that for each I in this range, the origin has a pair of symmetric homoclinic orbits. Interpreting these results in the full four-dimensional phase space, the set

$$\mathcal{M} = \{(x, y, I, \theta) \mid x = y = 0, \frac{k^2}{2} < I < I_{max}\} \tag{8.82}$$

is a two-dimensional manifold invariant under the flow generated by (8.80) (or, in other words, the vector field (8.80) is tangent to $\mathcal{M}$). In our definition of $\mathcal{M}$ we have restricted the range of I so that the fixed point at the origin in (8.81) is hyperbolic, and have further bounded the range of I values by choosing $I < I_{max}$ in order that $\mathcal{M}$ be bounded. (In practice, we just choose I_{max} large enough to contain the unperturbed structures of interest). These two requirements result in $\mathcal{M}$ being a *normally hyperbolic invariant manifold*.

In this integrable, four-dimensional system, homoclinic orbits connecting the origin in (8.81) are manifested as the coincidence of the three-dimensional stable and unstable manifolds of $\mathcal{M}$:

$$W^s(\mathcal{M}) \equiv W^u(\mathcal{M}) \equiv \{(x, y, I, \theta) | H(x, y, I) - H(0, 0, I) = 0, \ I > \frac{k^2}{2}\}. \tag{8.83}$$

It should be clear that trajectories in (8.83) approach a trajectory in $\mathcal{M}$ asymptotically as $t \to \pm\infty$. In Figure 8.17 we illustrate the invariant manifold structure that will be important for our analysis.

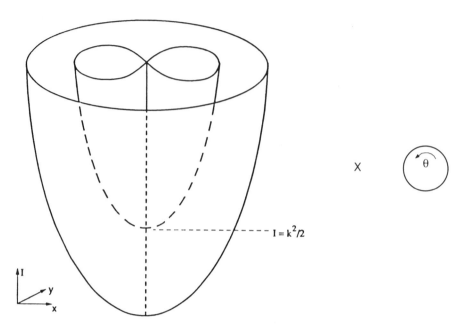

Figure 8.17. The invariant manifold structure that plays a role in our analysis. The outer paraboloid represents the boundary on which the variables (x, y, I, θ) become singular.

The Dynamics on $\mathcal{M}$

The unperturbed vector field restricted to $\mathcal{M}$ is given by

$$
\begin{aligned}
\dot{I} &= 0, \\
\dot{\theta} &= 1 - I.
\end{aligned}
\tag{8.84}
$$

The dynamics described by (8.84) are quite simple: all trajectories lie on periodic orbits, *except at* $I = 1$. At $I = 1$ the frequency $(\dot{\theta})$ vanishes, which results in a circle of fixed points. Thus we have a *resonance* and we will often refer to $I = 1$ as the *resonant I level* or *value*.

The Unperturbed Homoclinic Orbits and Their Relationship to the Dynamics on $\mathcal{M}$ and $W^s(\mathcal{M}) \cap W^u(\mathcal{M})$

In calculating the Melnikov functions, it will be important to have analytical expressions for the homoclinic orbits for (8.80), that is, solutions of (8.80) which are homoclinic to the invariant plane $\mathcal{M}$. These orbits can be constructed from solutions of (8.81) which are homoclinic to the origin $(x, y) = (0, 0)$. The calculations require a fair amount of tedious labor; however, they are instructive, so we include some of the details here.

We begin by defining the following coordinate change:

$$
x + iy = \sqrt{2B}\, e^{i\phi}.
\tag{8.85}
$$

In these coordinates, the unperturbed equations (8.80) become

$$\dot{B} = -2B(I - B)\sin 2\phi,$$

$$\dot{\phi} = k^2 - I(1 + \cos 2\phi) + B\left(\frac{3}{2} + 2\cos 2\phi\right),$$

$$\dot{I} = 0,$$

$$\dot{\theta} = 1 - I - B(1 + \cos 2\phi), \qquad (8.86)$$

with the Hamiltonian H being transformed into

$$H = \frac{1}{2}I^2 - I - \frac{3}{4}B^2 + (I - k^2)B + B(I - B)\cos 2\phi. \qquad (8.87)$$

By continuity, the value of the Hamiltonian on the orbits homoclinic to $\mathcal{M}$ is the same as its value on the "target" $\mathcal{M}$; specifically, $H = \frac{1}{2}I^2 - I$. Equating this value to (8.87) and canceling the common factor B gives

$$B = I - \frac{4k^2 - I}{3 + 4\cos 2\phi}. \qquad (8.88)$$

Substituting (8.88) into the $\dot{\phi}$ component of (8.86) gives the following equation for $\dot{\phi}$ on the homoclinic orbits:

$$\dot{\phi} = I(1 + \cos 2\phi) - k^2. \qquad (8.89)$$

Next we let

$$\psi = \theta + \phi \qquad (8.90)$$

and add (8.89) and the $\dot{\theta}$ component of (8.86) to get the following equation for $\dot{\psi}$ on the homoclinic orbits:

$$\dot{\psi} = 1 - I - \frac{1}{4}B. \qquad (8.91)$$

Equations (8.88), (8.89), and (8.91) provide us with the necessary (and sufficiently simple) relationships for solving for the homoclinic orbits. As one integrates, there are two cases corresponding to $I > 4k^2$ and $\frac{k^2}{2} < I < 4k^2$. For our study of the two-mode model we will only need the latter. Complete expressions for the homoclinic orbits can be found in Kovačič and Wiggins [1992].

$$\boxed{\frac{k^2}{2} < I < 4k^2 \text{ with initial conditions } \phi(t = 0) = 0, \ \psi(t = 0) = \theta_0.}$$

For this case we have

$$B = \frac{4\lambda^2}{(4k^2 - I)\cosh(2\lambda\,t) + (3k^2 + I)}, \qquad (8.92)$$

$$\tan\phi_{\pm} = \frac{\lambda}{k^2}\tanh(\lambda t), \qquad (8.93)$$

$$\psi = -\frac{1}{\sqrt{7}} \tanh^{-1} \left[\frac{\lambda}{7k^2} \tanh(\lambda t) \right] + (1 - I)t + \theta_0, \qquad (8.94)$$

where

$$\lambda = \lambda(k, I) \equiv k\sqrt{2I - k^2},$$

$$\phi_+ \in (\frac{-\pi}{2}, \frac{\pi}{2}); \quad \phi_- \in (\frac{\pi}{2}, \frac{3\pi}{2}).$$

Orbits Homoclinic to the Circle of Fixed Points

The unperturbed equations restricted to the invariant plane are given

$$\dot{I} = 0,$$
$$\dot{\theta} = 1 - I; \qquad (8.95)$$

hence, all orbits on the plane $\mathcal{M}$ are concentric circles centered at the origin. Each homoclinic orbit given by (8.92), (8.93), and (8.94) is homoclinic to one of these concentric circles in the annulus $\mathcal{M}$. Of these circles, the one with radius $I = 1$ is a circle of fixed points on $\mathcal{M}$. At $I = 1$, the homoclinic orbits are actually heteroclinic to fixed points on this circle. We use the expressions (8.92), (8.93), and (8.94) for the homoclinic orbits to compute the phase shift, $\Delta\theta$, of orbits that are asymptotic to points on the circle of fixed points as $t \to \pm\infty$.

$$\boxed{\tfrac{1}{2} < k < \sqrt{2}}$$

For this case, using (8.93) and (8.94) at $I = 1$ gives

$$\psi(-\infty) = \frac{1}{\sqrt{7}} \tanh^{-1} \left(\sqrt{\frac{2 - k^2}{7k^2}} \right) + \psi_0, \qquad (8.96)$$

$$\psi(\infty) = -\frac{1}{\sqrt{7}} \tanh^{-1} \left(\sqrt{\frac{2 - k^2}{7k^2}} \right) + \psi_0, \qquad (8.97)$$

$$\phi(-\infty) = -\tan^{-1} \left(\frac{\sqrt{2 - k^2}}{k} \right), \qquad (8.98)$$

$$\phi(\infty) = \tan^{-1} \left(\frac{\sqrt{2 - k^2}}{k} \right). \qquad (8.99)$$

Using (8.90) gives

$$\theta(-\infty) = \theta_0 + \tan^{-1} \left(\frac{\sqrt{2 - k^2}}{k} \right) + \frac{1}{\sqrt{7}} \tanh^{-1} \left(\sqrt{\frac{2 - k^2}{7k^2}} \right), \qquad (8.100)$$

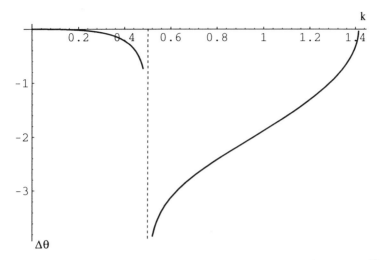

Figure 8.18. The graph of $\Delta\theta$ as a function of k for $\frac{1}{2} < k < \sqrt{2}$.

$$\theta(\infty) = \theta_0 - \tan^{-1}\left(\frac{\sqrt{2 - k^2}}{k}\right) - \frac{1}{\sqrt{7}}\tanh^{-1}\left(\sqrt{\frac{2 - k^2}{7k^2}}\right). \tag{8.101}$$

Hence, we have

$$\Delta\theta \equiv \theta(+\infty) - \theta(-\infty) = -2\tan^{-1}\left(\frac{\sqrt{2 - k^2}}{k}\right) - \frac{2}{\sqrt{7}}\tanh^{-1}\left(\sqrt{\frac{2 - k^2}{7k^2}}\right). \tag{8.102}$$

In Figure 8.18 we display $\Delta\theta$ as a function of k for this case. From this figure we see that $\Delta\theta$ is a monotonic function that approaches $-\infty$ as k goes to $\frac{1}{2}$ and zero as k goes to $\sqrt{2}$. Note that the function is always negative.

8.5.2.   The Perturbed Structure.

The Dynamics on $\mathcal{M}_\epsilon$, Near Resonance:Non-Hamiltonian Perturbations

The perturbed vector field (8.76) restricted to $\mathcal{M}_\epsilon$ is given by

$$\begin{aligned} \dot{I} &= -\epsilon\left(\Gamma\sqrt{2I}\cos\theta + 2\alpha I\right), \\ \dot{\theta} &= 1 - I + \frac{\epsilon\Gamma}{\sqrt{2I}}\sin\theta. \end{aligned} \tag{8.103}$$

We will be interested in the dynamics of (8.103) near $I = 1$, i.e., in an annulus on the cylinder with coordinates $I - \theta$ centered at $I = 1$ and we emphasize that *all questions concerning stability and dimensions of stable and unstable manifolds addressed in this subsection are made with respect to this two-dimensional dynamical system on* $\mathcal{M}_\epsilon$. Following Kovačič and Wiggins [1992], we rewrite (8.103) in a form suited for this purpose. We introduce the coordinate transformation

$$I = 1 + \sqrt{\epsilon\sqrt{2}\Gamma}\, h, \tag{8.104}$$

rescale time by letting

$$\tau = \sqrt{\epsilon\sqrt{2}\Gamma}\, t, \tag{8.105}$$

combine the parameters as follows

$$\eta \equiv \sqrt{\epsilon\sqrt{2}\Gamma}, \qquad \chi_\alpha \equiv \sqrt{2}\frac{\alpha}{\Gamma},$$

and Taylor expand in η about $\eta = 0$. The equations (8.103) then become

$$
\begin{aligned}
h' &= -\cos\theta - \chi_\alpha - \eta\left(\chi_\alpha + \frac{1}{2}\cos\theta\right)h + \mathcal{O}(\eta^2), \\
\theta' &= -h + \frac{\eta}{2}\sin\theta + \mathcal{O}(\eta^2),
\end{aligned}
\tag{8.106}
$$

where "$'$" denotes differentiation with respect to τ. The $h - \theta$ coordinates localize our study to a region near the resonance. Since we will be interested mainly in the dynamics near the resonance we will restrict the domain of $\mathcal{M}_\epsilon$ to an annulus containing the resonance. More precisely, the region of interest on $\mathcal{M}_\epsilon$ is defined as follows:

$$\mathcal{A} = \{(x, y, h, \theta) \,|\, x = y = 0, |h| \le C\}, \tag{8.107}$$

where C is an $\mathcal{O}(1)$ constant chosen sufficiently large to contain the resonance structures, i.e. the relevant segments of $W^u(q_\epsilon)$ and $W^s(q_\epsilon)$.

For $\eta = 0$, the equations reduce to

$$
\begin{aligned}
h' &= -\cos\theta - \chi_\alpha, \\
\theta' &= -h.
\end{aligned}
\tag{8.108}
$$

These equations are Hamiltonian with Hamiltonian function given by

$$\mathcal{H} = \frac{h^2}{2} - \sin\theta - \chi_\alpha\theta, \tag{8.109}$$

which is just the familiar Hamiltonian for a pendulum subject to a constant force. A simple analysis shows that equation (8.108) has two fixed points whose coordinates must satisfy

$$
\begin{aligned}
h &= 0, \\
\cos\theta &= -\chi_\alpha.
\end{aligned}
\tag{8.110}
$$

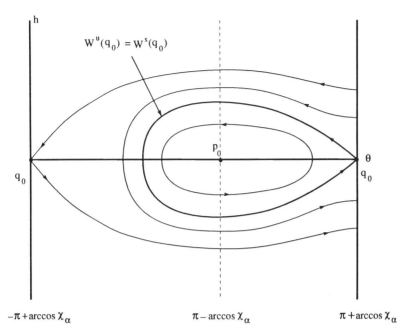

Figure 8.19. The structure of the phase portrait of the leading order approximate Hamiltonian equations for $0 < \chi_\alpha < 1$.

Hence, the (h, θ) coordinates of the two fixed points are given by

$$
\begin{aligned}
p_0 &= (h_{p_0}, \theta_{p_0}) = (0, \pi - \arccos \chi_\alpha), \\
q_0 &= (h_{q_0}, \theta_{q_0}) = (0, \pi + \arccos \chi_\alpha),
\end{aligned}
\tag{8.111}
$$

where p_0 is a center and q_0 is a saddle. These two fixed points coalesce in a Hamiltonian saddle-node bifurcation at $\chi_\alpha = 1$. Moreover, the saddle point is connected to itself by a homoclinic orbit. In Figure 8.19 we show the qualitative structure of the phase portrait of (8.108) for $0 < \chi_\alpha < 1$, and we will henceforth refer to this structure consisting of the homoclinic connection and its interior as the "fish". We denote the "nose of the fish", (i.e., the θ coordinate of the intersection of the separatrix with the θ axis) by θ_n. Note that the length (measured with respect to the θ coordinate) of the "fish" in Figure 8.19 varies monotonically with χ_α (from 0 to 2π). Moreover, the dynamics on $\mathcal{M}_\epsilon$ do not depend on k or β.

Next, we determine what becomes of the phase portrait when the higher order terms in η are taken into account. This is particularly easy in this case since the trace of the linearization of (8.106) is constant through $\mathcal{O}(\eta)$ and is given by $-\eta \chi_\alpha + \mathcal{O}(\eta^2)$. Hence, from an application of the implicit function theorem and standard phase plane results (Wiggins [1990]), for η sufficiently small and $0 < \chi_\alpha < 1$, p_0 becomes a sink, denoted p_ϵ, q_0 remains a saddle, denoted q_ϵ, and the homoclinic orbit breaks with a branch of the unstable manifold of q_ϵ falling into p_ϵ as shown in Figure 8.20.

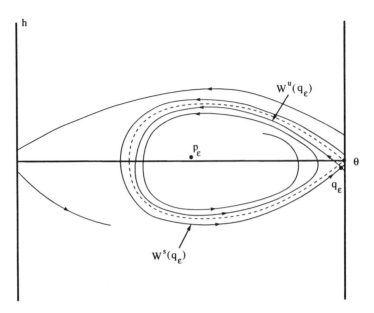

Figure 8.20. An approximate phase portrait for a typical value of Γ and α.

The Dynamics on $\mathcal{M}_\epsilon$, Near Resonance: Hamiltonian Perturbations

Using the same scalings as for the non-Hamiltonian case, the Hamiltonian of the reduced system is obtained by setting $\chi_\alpha = 0$,

$$\mathcal{H} = \frac{h^2}{2} - \sin\theta, \qquad (8.112)$$

with corresponding equations of motion

$$\begin{aligned} h' &= -\cos\theta, \\ \theta' &= -h. \end{aligned} \qquad (8.113)$$

Notice that (8.113) is just the equation for the familiar simple pendulum. This system has a hyperbolic fixed point at

$$q_0 = \left(0, \frac{\pi}{2}\right),$$

and an elliptic fixed point at

$$p_0 = \left(0, \frac{3\pi}{2}\right).$$

The hyperbolic fixed point is connected to itself by a pair of homoclinic orbits. (See Figure 8.21.) We now fix some $h_0 > 2\sqrt{\sqrt{2}\Gamma}$, so that all the rotational orbits and separatrices of the pendulum are internal orbits, in accordance with our earlier discussion.

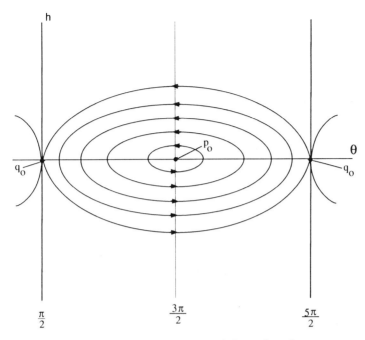

Figure 8.21. The phase portrait of the reduced system.

Explicit Evaluation of the Melnikov Function at $I = 1$

We now describe some details of the explicit computation of the Melnikov function. For our measurements, the value $I = 1$ will be sufficient, and we restrict the evaluation to this case. In the course of our calculation we will not need to specify θ_0. This means that the result can be equally applied when we study orbits homoclinic to either p_ϵ or q_ϵ; we will just need to choose the appropriate value of θ_0 for the case of interest.

The integrand of the Melnikov function is evaluated on an unperturbed homoclinic orbit, and, after some algebraic manipulation, can be shown to be

$$
\begin{aligned}
\langle n, (g^x, g^I, g^\theta) \rangle \;=\; & -\frac{dH_1}{dt}(x, y, I, \theta) - \frac{\partial H_1}{\partial \theta}(x, y, I, \theta)\frac{\partial H}{\partial I}(x, y, I) \\[2mm]
& -\; \beta\,(y\dot{x} - x\dot{y}) + \frac{\partial H_1}{\partial \theta}(x, y, I, \theta)\left(\frac{\partial H}{\partial I}(x, y, I) - \frac{\partial H}{\partial I}(0, 0, I = 1) \right) \\[2mm]
& +\; 2\alpha I\dot{\theta} + (\beta - \alpha)\left(x^2 + y^2\right)\dot{\theta},
\end{aligned}
\tag{8.114}
$$

where H is given by (8.78) and H_1 by (8.79),

$$
H_1 \equiv -\Gamma\,\sqrt{2I - x^2 - y^2}\,\sin\theta,
$$

which denotes the Hamiltonian part of the perturbation that is due solely to the external driving. At $I = 1$, (8.114) simplifies to the following:

$$\frac{-dH_1}{dt} + \beta\left(x\dot{y} - y\dot{x}\right) + 2\alpha\dot{\theta} + (\beta - \alpha)\left(x^2 + y^2\right)\dot{\theta}. \tag{8.115}$$

We now integrate (8.115) around the unperturbed homoclinic orbit at $I = 1$. We examine each term in (8.115) individually.

It is clear that the first term in (8.115) can be integrated directly to give

$$-\int_{-\infty}^{+\infty} \frac{dH_1}{dt} dt = \sqrt{2}\Gamma\left[\sin\theta(+\infty) - \sin\theta(-\infty)\right]. \tag{8.116}$$

It is also easy to see that the third term in (8.115) can be integrated directly to give

$$2\alpha \int_{-\infty}^{+\infty} \dot{\theta} dt = 2\alpha\Delta\theta. \tag{8.117}$$

We now examine the second and fourth terms in (8.115). Differentiating (8.85) with respect to time, one can easily show that

$$x\dot{y} - y\dot{x} = 2B\dot{\phi} \tag{8.118}$$

and taking the modulus of (8.85) gives

$$x^2 + y^2 = 2B. \tag{8.119}$$

From these two relations and (8.90) we obtain

$$\dot{\phi} = \dot{\psi} - \dot{\theta} = \frac{x\dot{y} - y\dot{x}}{x^2 + y^2}, \tag{8.120}$$

or

$$\dot{\theta} = \dot{\psi} - \frac{x\dot{y} - y\dot{x}}{x^2 + y^2}. \tag{8.121}$$

Substituting (8.121) into the fourth term of (8.115) and combining the result with the second term of (8.115) gives

$$\beta(x\dot{y} - y\dot{x}) + (\beta - \alpha)(x^2 + y^2)\dot{\theta}$$
$$= \alpha(x\dot{y} - y\dot{x}) + (\beta - \alpha)(x^2 + y^2)\dot{\psi}. \tag{8.122}$$

We next integrate the two terms in (8.122) around the homoclinic orbit. Using (8.118), the integral of the first term of (8.122) becomes

$$\int_{-\infty}^{+\infty} (x\dot{y} - y\dot{x}) dt = 2\int_{-\infty}^{+\infty} B\dot{\phi} dt. \tag{8.123}$$

Using (8.88), at $I = 1$ we have

$$B = 1 - (4k^2 - 1)\frac{1}{3 + 4\cos\phi},$$

which, when substituted into (8.123) gives

$$\int_{-\infty}^{+\infty} (x\dot{y} - y\dot{x})\, dt = 2\Delta\phi - 2(4k^2 - 1)\int_{\phi(-\infty)}^{\phi(+\infty)} \frac{d\phi}{3 + 4\cos\phi}, \tag{8.124}$$

where

$$\Delta\phi \equiv \phi(+\infty) - \phi(-\infty)$$

can be obtained from (8.98) and (8.99). Integrating the last term in (8.124) gives

$$\int_{-\infty}^{+\infty} (x\dot{y} - y\dot{x})\, dt = 2\Delta\phi - 2(4k^2 - 1)\Delta\psi, \tag{8.125}$$

where $\Delta\psi \equiv \psi(+\infty) - \psi(-\infty)$ and from (8.96), and (8.97) we have

$$\Delta\psi = -\frac{2}{\sqrt{7}}\tanh^{-1}\sqrt{\frac{2 - k^2}{7k^2}}, \quad \text{for} \quad \frac{1}{2} < k < \sqrt{2}. \tag{8.126}$$

Finally, we integrate the last term in (8.122) by first using (8.119) to obtain

$$\int_{-\infty}^{+\infty} (x^2 + y^2)\dot{\psi}\, dt = -\frac{1}{2}\int_{-\infty}^{+\infty} B^2\, dt. \tag{8.127}$$

Using the expressions for $B(t)$ in (8.92) gives

$$-\frac{1}{2}\int_{-\infty}^{+\infty} B^2\, dt = \frac{8}{7}\left[k\sqrt{2 - k^2} + (1 + 3k^2)\Delta\psi\right], \tag{8.128}$$

where $\Delta\psi$ is given in (8.126). Using (8.116), (8.117), (8.122), (8.125), (8.127), and (8.128), the Melnikov function at $I = 1$ becomes

$$M(1, \theta(-\infty), \chi_\alpha, \chi_\beta, k) = \sqrt{2}\Gamma \left\{ [\sin\theta\,(+\infty) - \sin\theta\,(-\infty)] + 4\chi_\alpha k^2 \Delta\psi \right.$$
$$\left. + (\chi_\beta - \chi_\alpha)\frac{4}{7}\left[k\sqrt{2 - k^2} + (1 + 3k^2)\Delta\psi\right] \right\}, \tag{8.129}$$

where

$$\theta(+\infty) = \theta(-\infty) + \Delta\theta,$$
$$\chi_\alpha = \sqrt{2}\frac{\alpha}{\Gamma},$$
$$\chi_\beta = \sqrt{2}\frac{\beta}{\Gamma}. \tag{8.130}$$

Following the discussion in Section 4.5 we have the interpretation

$$\theta(-\infty) \equiv \theta_b, \tag{8.131}$$
$$\theta(+\infty) \equiv \theta_b^L = \theta_b + \Delta\theta. \tag{8.132}$$

8.5.3.   Orbits Homoclinic to p_ϵ: Non-Hamiltonian Perturbations. Following the theory developed earlier, in order to show that there exists an orbit homoclinic to p_ϵ, we must first show that the Melnikov function evaluated on the unperturbed orbit that is backwards asymptotic to p_0 has a simple zero. This means taking $\theta_b = \pi - \arccos \chi_\alpha$ in the expression for the Melnikov function given in (8.129). This simple zero of the Melnikov function is a sufficient condition for the existence of an orbit that is asymptotic to p_ϵ as $t \to -\infty$ and asymptotic to an orbit in $\mathcal{A}_\epsilon$ as $t \to +\infty$. In order to verify that this orbit is asymptotic to an orbit in $\mathcal{A}_\epsilon$ that approaches p_ϵ as $t \to +\infty$, it is sufficient to show that the unperturbed heteroclinic orbit that is asymptotic to p_0 as $t \to -\infty$ returns to the circle of fixed points as $t \to +\infty$ at a θ value that places it within the unperturbed homoclinic orbit in $\mathcal{A}_\epsilon$ (in the rescaled coordinates) that connects q_0. In order to simplify the calculations, we take $k = 1$. *Also, we note that in the following we round all numerical quantities (i.e., $\Delta\theta$ and $\Delta\psi$ at $k = 1$) to two decimal places. In practice, these can be computed to any desired degree of accuracy; however, recall that the relevant quantities are only $\mathcal{O}(\epsilon)$ or $\mathcal{O}(\sqrt{\epsilon})$ approximations to the actual quantities for the perturbed system.*

Setting $k = 1$, and using (8.102), (8.126), and (8.129), the condition that $M(\chi_\alpha, \chi_\beta, k = 1) = 0$ is given by

$$\chi_\beta = -11.20\sqrt{1 - \chi_\alpha^2} - 1.14\chi_\alpha. \tag{8.133}$$

Thus it follows that for all values of $\chi_\alpha \in (0,1)$, there is a value of χ_β (which is always negative) such that the Melnikov function is zero. Moreover, it is an easy calculation to show that this zero is a simple zero.

Next, we show that the unstable manifold of p_ϵ returns to $\mathcal{A}_\epsilon$ at the appropriate location. From (8.102), at $k = 1$ we have

$$\Delta\theta = -1.87. \tag{8.134}$$

The location of the nose of the fish is given by the solution of the following transcendental equation:

$$\mathcal{H}(0, \theta_n) - \mathcal{H}(0, \pi + \arccos \chi_\alpha)$$
$$= \sin(\pi + \arccos \chi_\alpha) - \sin(\theta_n) + \chi_\alpha(\pi + \arccos \chi_\alpha - \theta_n) = 0, \tag{8.135}$$

where $\mathcal{H}$ is given by (8.109) and $\chi_\alpha = \sqrt{2}\frac{\alpha}{\Gamma}$. From our knowledge of the phase portrait of the pendulum with torque, we know that (8.135) has exactly two solutions, one being the θ value corresponding to q_0, denoted θ_{q_0}. This equation can easily be solved numerically using Newton's method. In Figure 8.22 we present θ_n, θ_{q_0}, and $\theta_{p_0} + \Delta\theta$ for values of χ_α between 0 and 1. (Note: θ_{p_0} is the θ coordinate of p_0.)

Recall that $\Delta\theta$ only depends on k; in particular, it does not depend on χ_α or χ_β. Also, θ_{q_0} and θ_{p_0} depend only on χ_α. These facts allow for the simple graphical depiction in Figure 8.22 of the criteria for the unstable manifold of p_ϵ to asymptote to an orbit inside the "fish" on $\mathcal{A}_\epsilon$. Thus, we see that for $\chi_\alpha \in (0, 0.19)$ we have $\theta_n < \theta_{p_0} + \Delta\theta < \theta_{q_0}$. Hence, for ϵ sufficiently small,

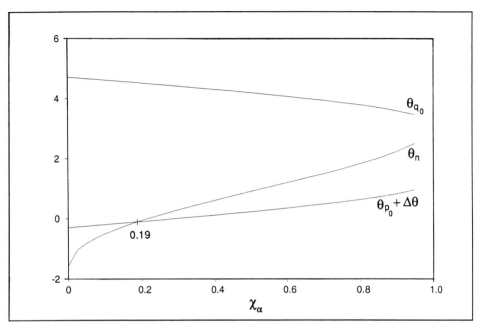

Figure 8.22. Graphs of θ_{q_0}, θ_n, and $\theta_{p_0} + \Delta\theta$.

$$\frac{\alpha}{\Gamma} < .13,$$

and

$$\frac{\beta}{\Gamma} \approx -7.92\sqrt{1 - 2(\tfrac{\alpha}{\Gamma})^2} - 1.14\frac{\alpha}{\Gamma},$$

p_ϵ has a symmetric pair of homoclinic orbits.

The "approximate" sign refers to the fact that the Melnikov function is only an accurate measure of distance up to $\mathcal{O}(\epsilon^2)$ and the numerical quantities have only been given to two decimal places. In Figure 8.23, we show the curve near which the homoclinic orbit occurs.

8.5.4.   Orbits Homoclinic to q_ϵ: Non-Hamiltonian Perturbations. The Melnikov function for this case is given in (8.129) by

$$M(1, \theta_b; \chi_\alpha, \chi_\beta, k) = \sqrt{2}\Gamma \left\{ \sin\theta_b^L - \sin\theta_b + 4\chi_\alpha k^2 \Delta\psi \right.$$
$$\left. + (\chi_\beta - \chi_\alpha)\frac{4}{7}\left(k\sqrt{2 - k^2} + (1 + 3k^2)\Delta\psi \right) \right\}, \qquad (8.136)$$

where

$$\theta_b^L \equiv \theta_b + \Delta\theta. \qquad (8.137)$$

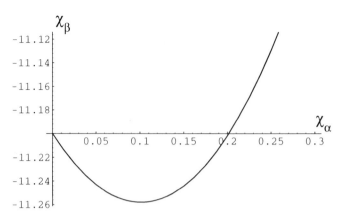

Figure 8.23. Parameter values in the $\frac{\beta}{\Gamma} - \frac{\alpha}{\Gamma}$ plane (with $k = 1$) for which orbits homoclinic to p_ϵ exist, $\frac{\alpha}{\Gamma} < .13$.

The energy difference defined in (8.67) is given by

$$\Delta\mathcal{H} = -\chi_\alpha \Delta\theta - \sin\theta_b^L + \sin\theta_b. \qquad (8.138)$$

Now we need only check that it is possible at a zero of the Melnikov function for $\mathcal{H}$ to take values throughout an $\mathcal{O}(1)$ interval about $\mathcal{H} = 0$. At a zero of the Melnikov function, equation (8.136) can be rearranged to yield

$$\sin\theta_b^T - \sin\theta_b^L = \tilde{A}(k)\chi_\alpha + B(k)\chi_\beta, \qquad (8.139)$$

where

$$\tilde{A}(k) \equiv \frac{4}{7}\left((4k^2 - 1)\Delta\psi - k\sqrt{2 - k^2}\right),$$

$$B(k) \equiv \frac{4}{7}\left((3k^2 + 1)\Delta\psi + k\sqrt{2 - k^2}\right), \qquad (8.140)$$

and $\Delta_\psi = \Delta_\psi(1, k)$. Using expression (8.139) to eliminate $\sin\theta_b^T - \sin\theta_b^L$ from (8.138), one finds that *at a zero of the Melnikov function,*

$$\Delta\mathcal{H} = \left(\tilde{A}(k) - \Delta\theta\right)\chi_\alpha + (B(k))\chi_\beta. \qquad (8.141)$$

Thus, *at a zero of the Melnikov function,* $\Delta\mathcal{H}$ is linear in χ_α and χ_β, with coefficients depending only upon k. Clearly, $\Delta\mathcal{H}$ can be made to oscillate between $[-O(1), +O(1)]$, for example by varying the parameter χ_β. Hence, q_ϵ has a symmetric pair of homoclinic orbits near the parameter values defined by

$$\left(\tilde{A}(k) - \Delta\theta\right)\chi_\alpha + (B(k))\chi_\beta = 0.$$

In McLaughlin *et al.* [1993] it is shown how breaking these homoclinic orbits gives rise to a Smale horseshoe type construction.

Remark. It is interesting to note that if one thinks of χ_α as damping in the complex c mode and χ_β as damping in the complex b mode, that orbits homoclinic to p_ϵ require χ_β negative and orbits homoclinic to q_ϵ require χ_β positive, for χ_α positive in both cases.

8.5.5. **Orbits Homoclinic to Internal Orbits: Hamiltonian Perturbations.** We are now in a position to apply the theory for Hamiltonian perturbations and determine the existence of orbits homoclinic to the periodic orbits inside the resonance as well as orbits homoclinic to the saddle fixed point. This amounts to applying the energy-phase criterion as formulated in Theorem 8.4.4. A simple calculation gives

$$\triangle\mathcal{H}(\theta;\Gamma,k) = \mathcal{H}(h,\theta;\Gamma) - \mathcal{H}(h,\theta - \triangle\theta(k);\Gamma) = 0 \quad \Rightarrow \quad \sin\theta - \sin(\theta - \triangle\theta) = 0. \tag{8.142}$$

It is easy to see that (8.142) has two solutions given by

$$\theta_1 = \frac{\triangle\theta}{2} + \frac{\pi}{2}, \tag{8.143}$$

and

$$\theta_2 = \frac{\triangle\theta}{2} + \frac{3\pi}{2}. \tag{8.144}$$

Using (8.12), we can show that for $\triangle\theta(k) = 0, 2\pi$ we have $D_\theta\triangle\mathcal{H}(\theta;\Gamma,k) = 0$ at $\theta_1 = \frac{\pi}{2}, \frac{3\pi}{2}$ and $\theta_2 = \frac{\pi}{2}, \frac{3\pi}{2}$, i.e. $D_\theta\triangle\mathcal{H}$ is zero at the θ values of the hyperbolic and elliptic fixed points in the resonance. According to Figure 8.1, this means that for $\triangle\theta \in [0, 2\pi]$, there are *two* values of k for which $\triangle\mathcal{H}(\theta;\Gamma,k) = 0$ and $D_\theta\triangle\mathcal{H}(\theta;\Gamma,k) = 0$. Using the θ-periodicity of $\triangle\mathcal{H}(\theta;\Gamma,k)$, it follows that there is a set Λ of countable infinity of k values (converging to $k = \frac{1}{2}$ from above and below) for which $D_\theta\triangle\mathcal{H}(\theta;\Gamma,k) = 0$.

Using (8.143) and (8.144) we have

$$Z_{\mu,1}^+ = \{(h,\theta)|\theta = \frac{\triangle\theta(k)}{2} + \frac{\pi}{2}\},$$

$$Z_{\mu,1}^- = \{(h,\theta)|\theta = \frac{\pi}{2} - \frac{\triangle\theta(k)}{2}\},$$

$$Z_{\mu,2}^+ = \{(h,\theta)|\theta = \frac{\triangle\theta(k)}{2} + \frac{3\pi}{2}\},$$

$$Z_{\mu,2}^- = \{(h,\theta)|\theta = \frac{3\pi}{2} - \frac{\triangle\theta(k)}{2}\},$$

defined for $k \notin \Lambda$.

For $\triangle\theta = 0 \bmod 2\pi$ $(k \in \Lambda)$ $Z_{\mu,1}^+$ and $Z_{\mu,1}^-$ are not defined but would be located at $\theta = \frac{\pi}{2}, \frac{5\pi}{2}$. As $\triangle\theta$ increases, $Z_{\mu,1}^+$ moves from $\theta = \frac{\pi}{2}$ monotonically towards $\frac{3\pi}{2}$ (with θ increasing) and $Z_{\mu,1}^-$ moves monotonically towards $\frac{3\pi}{2}$ (with θ decreasing), reaching these points at $\triangle\theta = 2\pi$, where they again cease to be defined.

Similarly, for $\Delta\theta = 0 \bmod 2\pi$ $(k \in \Lambda)$ $Z^+_{\mu,2}$ and $Z^-_{\mu,2}$ are not defined but would lie at $\frac{3\pi}{2}$. As $\Delta\theta$ increases, $Z^+_{\mu,2}$ moves monotonically from $\theta = \frac{3\pi}{2}$ towards $\frac{5\pi}{2}$ (with θ increasing) and $Z^-_{\mu,2}$ moves monotonically towards $\frac{\pi}{2}$ (with θ decreasing), reaching these points at $\Delta\theta = 2\pi$, where they cease to be defined. Also, at $\Delta\theta = \pi$, $Z^+_{\mu,2}$ coincides with $Z^-_{\mu,1}$ and $Z^+_{\mu,1}$ coincides with $Z^-_{\mu,2}$. This variation of the sets $Z^\pm_{\mu,1}$ and $Z^\pm_{\mu,2}$ repeats itself more and more rapidly as the parameter k approaches $1/2$. (See Figure 8.1.)

In Figures 8.24 to 8.26 we plot the location of $Z^+_{\mu,1}$, $Z^-_{\mu,1}$, $Z^+_{\mu,2}$, and $Z^-_{\mu,2}$ for a value of $\Delta\theta$ between 0 and π, for $\Delta\theta = \pi$, and for a value of $\Delta\theta$ between π and 2π, respectively. These three plots also indicate the main qualitatively different behaviors possible. We discuss each situation individually, but first, we want to introduce some terminology. As vertical lines, $Z^+_{\mu,1}$ and $Z^-_{\mu,1}$ are tangent to the right and left, respectively, extremal (in θ) points of a unique periodic orbit. All periodic orbits surrounding this orbit are intersected by $Z^+_{\mu,1}$ and $Z^-_{\mu,1}$ in two unique points, respectively (the homoclinic orbits are intersected in one point). The periodic orbits inside this periodic orbit do not intersect $Z^+_{\mu,1}$ and $Z^-_{\mu,1}$. We refer to the region outside this special periodic orbit as the *1-accessible region*, denoted R_{1A}. Similarly, a *2-accessible region* can be defined using $Z^+_{\mu,2}$ and $Z^-_{\mu,2}$ and will be denoted R_{2A}.

In Figures 8.24 to 8.26 we indicate the 1-accessible and 2-accessible regions. From our earlier discussions, periodic orbits that are in both R_{1A} and R_{2A} have eight transverse homoclinic orbits, and periodic orbits that are neither in R_{1A} nor R_{2A} have no (simple) transverse homoclinic orbits. (The reason the numbers are 0-4-8, rather than 0-2-4, is because the symmetry of the system effectively doubles the number of transverse homoclinic orbits.) The saddle point (q_0) has 8 transverse homoclinic orbits. Thus, for $0 < \Delta\theta < \pi$ and $\pi < \Delta\theta < 2\pi$, the interior of the resonance (i.e., the interior of the region bounded by the transverse homoclinic orbits connecting q_0) is partitioned into three regions: one containing periodic orbits that have no transverse homoclinic orbits, one containing periodic orbits that have four transverse homoclinic orbits, and one containing periodic orbits that have eight transverse homoclinic orbits. In passing between these regions a *saddle-node bifurcation of homoclinic orbits* occurs as described in Haller and Wiggins [1993a].

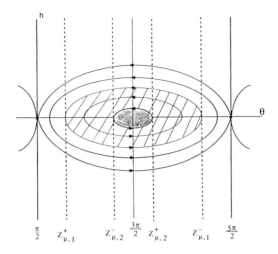

Figure 8.24. The energy-phase criterion for $\Delta\theta \in (0, \pi)$.

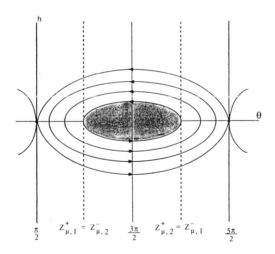

Figure 8.25. The energy-phase criterion for $\Delta\theta = \pi$.

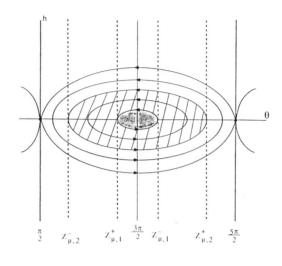

 0 Homoclinic Orbits

R_{1A} : 4 Homoclinic Orbits

R_{1A}, R_{2A} : 8 Homoclinic Orbits

Figure 8.26. The energy-phase criterion for $\Delta\theta \in (\pi, 2\pi)$.

References

Abraham, R. and Marsden, J.E. [1978] *Foundations of Mechanics*, The Benjamin/ Cummings Publishing Co. Inc.: Reading.

Arnold, V.I. [1962] On the Behavior of an Adiabatic Invariant Under Slow Periodic Variation of the Hamiltonian. *Sov. Math. Dokl.*, **3**, 136-139.

Arnold,V.I. [1964] Instability of Dynamical Systems with Many Degrees of Freedom. *Dokl. Akad. Nauk. USSR*, **156**, 9.

Arnold, V.I. [1963] Proof of A. N. Kolmogorov's Theorem on the Preservation of Quasiperodic Motions Under Small Perturbations of the Hamiltonian. *Russ. Math. Surveys*, **18**(5), 9-36.

Arnold,V.I., Kozlov,V.V., and Neishtadt, A.I. [1988] Mathematical Aspects of Classical and Celestial Mechanics in *Dynamical Systems III*, V.I. Arnold (ed.), Springer-Verlag: New York, Heidelberg, Berlin.

Ballal, B.Y. and Rivlin, R.S. [1976] Flow of a Newtonian Fluid Between Eccentric Rotating Cylinders: Inertial Effects. *Archiv. Rat. Mech. Anal.*, **62**, 237-294.

Bartlett, J. H. [1982] Limits of Stability for an Area Preserving Polynomial Mapping. *Cel. Mech.*, **28**, 295-317.

Beigie, D. Leonard, A. and Wiggins, S. [1991] Chaotic Transport in the Homoclinic and Heteroclinic Tangle Regions of Quasiperiodically Forced Two Dimensional Dynamical Systems. *Nonlinearity*, **4**, 775-819.

Beigie, D. and Wiggins, S. [1992] The Dynamics Associated with a Quasiperiodically Forced Morse Oscillator: Application to Molecular Dissociation. *Physical Review A*, **45**(7), 4803-4827.

Beigie, D., Leonard A., and Wiggins, S. [1992] The Dynamics Associated with the Chaotic Tangles of Two-Dimensional Quasiperiodic Vector Fields: Theory and Applications, in *Nonlinear Phenomena in Atmospheric and Oceanic Sciences* (IMA Volumes in Mathematics and its Applications - Volume 40). Editors G.F. Carnevale and R. Pierrehumbert (Springer-Verlag: New York, Berlin, Heidelberg) 47-138.

Bishop, A. R., Forest, M.G., McLaughlin, D.W. and Overman, E.A., II [1986] Coherence and Chaos in the Driven Damped Sine-Gordon Equation: Measurement of the Soliton Spectrum. *Physica D*, **23**, 293.

Bishop, A. R., McLaughlin, D.W., Forest, M.G. and Overman, E.A., II [1988] A Quasiperiodic Route to Chaos in a Near-Integrable PDE. *Physica D*, **23**, 293-328.

Bishop A. R., Forest, M.G., McLaughlin, D.W. and Overman, E.A., II [1990a] A Modal Representation of Chaotic Attractors for the Driven, Damped Pendulum Chain. *Phys. Lett. A*, **144**, 17-25.

Bishop, A. R., Flesch, R., Forest, M.G., McLaughlin, D.W. and Overman, E.A., II [1990b] Correlations Between Chaos in a Perturbed Sine-Gordon Equation and a Truncated Model System. *SIAM J. Math. Anal.*, **21**, 1511-1536.

Bolton, E.W., Busse, F. and Clever, R. M. [1986] Oscillatory Instabilities of Convection Rolls at Intermediate Prandtl Numbers. *J. Fluid Mech.*, **164**, 469-486.

Brown, G. L. and Roshko, A. [1974] On Density Effects and Large Structure in Mixing. *J. Fluid Mech.*, **64**, 775-816.

Bruhwiler, D.L. and Cary, J.R. [1989], Diffusion of Particles in a Slowly-Modulated Wave. *Physica D*, **40**, 265-282.

Camassa, R. and Wiggins, S. [1991] Chaotic Advection in a Rayleigh-Bénard Flow. *Phys. Rev. A*, **43**, 774-797.

Cary, J.R., Escande, D.F., and Tennyson, J. [1986] Adiabatic Invariant Change Due to Separatrix Crossing. *Phys. Rev. A*, **34**, no.5, 4256-4275.

Cary, J.R. and Skodje, R.T. [1989] Phase Change Between Separatrix Crossings. *Physica D*, **39** 287.

Chandrasekhar, S. [1961] *Hydrodynamics and Hydromagnetic Stability.* Dover: New York.

Channon, S. R. and Lebowitz, J. L. [1980] Numerical Experiments in Stochasticity and Homoclinic Oscillation. *Ann. New York Acad. Sci.*, **357**, 108-118.

Chiercia, L. and Gallavotti, G., Drift and Diffusion in Phase Space. Universitá di Roma, preprint (1992).

Chorin, A. J. and Marsden, J. E. [1979] *A Mathematical Introduction to Fluid Mechanics.* Springer-Verlag: New York, Heidelberg, Berlin.

Clever, R. H. and Busse, F. [1974] Transition to Time-Dependent Convection. *J. Fluid Mech.*, **65**, 625-645.

Conley, C. C. and Zehnder, E. [1983] The Birkhoff-Lewis Fixed Point Theorem and a Conjecture of V. I. Arnold. *Invent. Math.*, **73**, 33-49.

de la Llave, R. and Wayne, C.E. [1990] Whiskered and Low Dimensional Tori in Nearly Integrable Hamiltonian Systems. University of Texas, Austin, preprint.

Dellnitz, M., Melbourne, I. and Marsden, J.E. [1992] Generic Bifurcation of Hamiltonian Vector Fields with Symmetry. *Nonlinearity* **5**, 979–996.

Douady, R. and Le Calvez, P. [1983] Example de point fixe elliptique non topologiquement stable en dimension 4. *C. R. Acad. Sci. Paris, Sér. I.*, **296**, 895-898.

Eliasson, L.H. [1988] Perturbations of Stable Invariant Tori. *Ann. Sci. Norm. Super. Pisa Cl. Sci. IV.*, Ser. **15**, 115.

Elskens, Y. and Escande, D. [1991] Slowly Pulsating Separatrices Sweep Homoclinic Tangles where Islands Must be Small: An Extension of Classical Adiabatic Theory. *Nonlinearity*, **4**, 615.

Ercolani, N. Forest, M.G. and McLaughlin, D.W. [1990] Geometry of the Modulational Instability III: Homoclinic Orbits for the Periodic Sine-Gordon Equation. *Physica D*, **43**, 349.

Ercolani, N., Forest, M.G., and McLaughlin, D.W. Geometry of the Modulational Instability Part I: Local Analysis, to appear.

Ercolani, N., Forest, M.G., and McLaughlin, D.W., Geometry of the Modulational Instability Part II: Global Results, to appear.

Fenichel, N. [1971] Persistence and Smoothness of Invariant Manifolds for Flows. *Ind. Univ. Math. J.* **21**, 193-226.

Fenichel, N. [1974] Asymptotic Stability with Rate Conditions. *Ind. Univ. Math. J.* **23**, 1109-1137.

Fenichel, N. [1977] Asymptotic Stability with Rate Conditions II. *Ind. Univ. Math. J.* **26**, 81-93.

Fenichel, N. [1979] Geometric Singular Perturbation Theory for Ordinary Differential Equations. *J. Diff. Eqs.* **31**, 53.

Golubitsky, M. and Stewart, I. [1987] Generic Bifurcation of Hamiltonian Systems with Symmetry. *Physica D*, **24**, 391-405.

Haller, G. and Wiggins, S. [1993a] Orbits Homoclinic to Resonances: The Hamiltonian Case. *Physica D*, **66**, 298-346.

Haller, G. and Wiggins, S. [1993b] N-Pulse Homoclinic Orbits in Perturbations of Hyperbolic Manifolds of Hamiltonian Equilibria. Submitted to *Mathematische Zeitschrift*.

Hirsch, M.W., Pugh, C.C., and Shub, M. [1977] *Invariant Manifolds*. Lect. Notes. in Math. **583**, Springer-Verlag: New York, Heidelberg, Berlin.

Ide, K. and Wiggins, S. [1989] The Bifurcation to Homoclinic Tori in the Quasiperiodically Forced Duffing Oscillator (with K. Ide). *Physica D*,**34**, 169-182.

Kaper, T.J. [1991] Part I: On the Structure of Separatrix-Swept Regions in Slowly-Modulated Hamiltonian Systems; Part II: On the Quantification of Mixing in Chaotic Stokes Flows – The Eccentric Journal Bearing, PhD Thesis, California Institute of Technology.

Kaper, T.J. and Wiggins, S. [1992] On the Structure of Separatrix-Swept Regions in Singularly-Perturbed Hamiltonian Systems. *Differential and Integral Equations*, **5**(6), 1363-1381.

Kaper, T.J. and Wiggins, S. [1993] An Analytical Study of Transport in Stokes Flows Exhibiting Large Scale Chaos in the Eccentric Journal Bearing. *J. Fluid Mech.*, **253**, 211-243.

Kolmogorov, A. N. [1954] On Conservation of Conditionally Periodic Motions Under Small Perturbations of the Hamiltonian. *Dokl. Akad. Nauk. USSR*, **98**(4), 527-530.

Kovačič, G. and S. Wiggins [1992] Orbits Homoclinic to Resonances, with an Application to Chaos in a Model of the Forced and Damped Sine-Gordon Equation. *Physica D*, **57**, 185-225.

Kovačič, G [1989] Ph. D. Thesis, California Institute of Technology.

Leal, L.G. [1992] *Laminar Flow and Convective Transport Processes: Scaling Principles and Asymptotic Analysis.* Butterworth-Heinemann, Boston.

Lichtenberg, A. J. and Lieberman, M.A. [1982]. *Regular and Stochastic Motion.* Springer-Verlag: New York, Heidelberg, Berlin.

Lochak, P. [1992] Canonical Perturbation Theory via Simultaneous Approximation, D.M.I. preprint.

Lochak, P., Neishtadt, A. [1992] Estimates of Stability Time for Early Integrable Systems with a Quasiconvex Hamiltonian. *Chaos*, **4**(2), 495-500.

MacKay, R. S., Meiss, J. D. and Percival, I. C. [1984] Transport in Hamiltonian Systems. *Physica 13D* 55–81.

MacKay, R. S., Meiss, J. D. and Percival, I. C. [1987] Resonances in Area Preserving Maps. *Physica* 27D, 1–20.

MacKay, R. S. and Meiss, J. D. [1992] Cantori for Symplectic Maps Near the Anti-Integrable Limit. *Nonlinearity*, **5**(1), 149-160.

Marsden, J. E. [1992]*Lectures on Mechanics.* Cambridge University Press: Cambridge.

McLaughlin, D. Overman II, E. A., Wiggins, S. and Xiong, C. [1993] Homoclinic Orbits in a Four-Dimensional Model of a Perturbed NLS Equation: A Geometric Singular Perturbation Study. Submitted to *Dynamics Reported.*

Melnikov, V.K. [1963] On the Stability of the Center for Time Periodic Perturbations. *Trans. Mosc. Math. Soc.*, **12**, 1-57.

Meyer, K. [1974] Generic Bifurcations in Hamiltonian Systems, in: *Dynamical Systems: Warwick 1974*, A. Mannning, ed., Lecture Notes in Math., vol. 468. Springer-Verlag: New York, Heidelberg, Berlin.

Meyer, K.R. and Sell, G.R. [1989] Melnikov Transforms, Bernoulli Bundles, and Almost Periodic Perturbations. *Trans. Am. Math. Soc.* **314**, 63-105.

Moser, J. [1962] On Invariant Curves of an Area Preserving Mappings of an Annulus. *Nachr. Akad. Wiss. Gött., II. Math.-Phys. Kl.*, 1-20.

Neishtadt, A.I. [1975] Passage through a Separatrix in a Resonance Problem with a Slowly-Varying Parameter. *PMM*, **39**, no.4, 594-605.

Neishtadt, A.I., Chaikovskii, D.K. and Chernikov, A.A. [1991] Adiabatic Chaos and Particle Diffusion. *Sov. Phys. JETP.* **72**, no.3, 423-430.

Nekhorosev, N.N. [1977] An Exponential Estimate on the Time of Stabilty of Nearly-Integrable Hamiltonian Systems. *Russ. Math. Surv.* **32**, 1.

Ottino, J. [1989] *The Kinematics of Mixing: Stretching, Chaos, and Transport.* Cambridge University Press: Cambridge.

Palmer, K. [1986] Transversal Heteroclinic Points and Cherry's Example of a Non-integrable Hamiltonian System. *JDE*, **65**, 321-360.

Pöschel, J. [1989] On Elliptic Lower Dimensional Tori in Hamiltonian Systems. *Math. Z.*, **202**, 559.

Pöschel, J. [1993] Nekhoroshev Estimates for Quasi-Convex Hamiltonian Systems. To appear in *Mathematische Zeitschrift*.

Robinson, C. [1983] Sustained Resonance for a Nonlinear System with Slowly-Varying Coefficients. *SIAM Math Anal*, **14**, no.5, 847-860.

Rom-Kedar, V., Leonard, A. and Wiggins, S. [1990] An Analytical Study of the Transport, Mixing and Chaos in an Unsteady Vortical Flow. *J. Fluid Mech.*, **214**, 347-394.

Rom-Kedar, V. and Wiggins, S. [1990] Transport in Two-Dimensional Maps. *Archive for Rational Mechanics and Analysis*, **109**, 3, 239-298.

Salamon, D. [1990] Morse Theory, the Conley Index, and Floer Homology. *Bull. London Math. Soc.*, **22**, 113-140.

Scheurle, J. [1986] Chaotic Solutions of Systems with almost Periodic Forcing. *J. Appl. Math. and Phys.* (ZAMP), **37**, 12-26.

Solomon, T.H. and Gollub, J.P. [1988] Chaotic Particle Transport in Time-Dependent Rayleigh-Bénard Convection. *Phys. Rev. A*, **38**, 6280-6286.

Stoffer, D. [1988a] Transversal Homoclinic Points and Hyperbolic Sets for Non-Autonomous Maps I. *J. Appl. Math. and Phys.* (ZAMP), **39**, 518-549.

Stoffer, D. [1988b] Transversal Homoclinic Points and Hyperbolic Sets for Non-Autonomous Maps II. *J. Appl. Math. and Phys.* (ZAMP), **39**, 783-812.

Treshchev, D.V. [1991] The Mechanism of Destruction of Resonant Tori of Hamiltonian Systems. *Math. USSR Sb.*, **68**, 181.

Wiggins, S. [1988a] *Global Bifurcations and Chaos – Analytical Methods.* Springer-Verlag Applied Mathematical Science Series, second printing 1990.

Wiggins, S. [1988b] On the Detection and Dynamical Consequences of Orbits Homoclinic to Hyperbolic Periodic Orbits and Normally Hyperbolic Invariant Tori in a Class of Ordinary Differential Equations. *SIAM J. Appl. Math.*, **48**, 262-285.

Wiggins, S. [1988c] Adiabatic Chaos. *Phys. Lett. A.*, **128**, 339-342.

Wiggins, S. [1990] *Introduction to Applied Nonlinear Dynamical Systems and Chaos.* Springer-Verlag Texts in Applied Mathematics Series, second printing 1991.

Wiggins, S. [1992] *Chaotic Transport in Dynamical Systems.* Springer-Verlag Interdisciplinary Applied Mathematical Sciences Series.

Wiggins, S. [1993] *Normally Hyperbolic Invariant Manifolds in Dynamical Systems.* Springer-Verlag: New York, Heidelberg, Berlin.

Index